W0106943

ICME IV
The Fourth International Congress
on Mathematical Education

Teaching Teachers, Teaching Students

Reflections on Mathematical Education

Lynn Arthur Steen
Donald J. Albers,
editors

Springer Science+Business Media, LLC 1981

Editors

Lynn Arthur Steen
Department of Mathematics
St. Olaf College
Northfield, Minnesota 55057

Donald J. Albers
Mathematics Department
Menlo College
Menlo Park, California 94025

Library of Congress Cataloging in Publication Data
International Congress on Mathematical Education (4th : 1980 :
   University of California, Berkeley)
   Teaching teachers, teaching students.
   Bibliography: p.
   Includes index.
   1. Mathematics--Study and teaching-- Congresses.
I. Steen, Lynn Arthur, 1941-   .   II. Albers, Donald J., 1941-
III. Title.
QA11.A1I46   1980   510'.7'1   81-1324
ISBN 978-0-8176-3043-0   AACR2

CIP — Kurztitelaufnahme der Deutschen Bibliothek
Teaching teachers, teaching students : reflections on mathemat. educa-
tion / ICME IV, the 4. Internat. Congress on Mathemat. Education.
Ed. by Lynn Arthur Steen ; Donald J. Albers. — Boston ; Basel ;
Stuttgart : Birkhäuser, 1981.
ISBN 978-0-8176-3043-0   ISBN 978-1-4899-0427-0 (eBook)
DOI 10.1007/978-1-4899-0427-0
NE: Steen, Lynn Arthur (Hrsg.); International Congress on Mathematical
Education (04, 1980, Berkeley, Calif.)

All rights reserved. No part of this publication may be reproduced,
stored in a retireval system, or transmitted, in any form or by any
means, electronic, mechanical, photocopying, recording or otherwise,
without prior permission of the copyright owner.

© Springer Science+Business Media New York 1981
Originally published by Birkhäuser Boston in 1981
Softcover reprint of the hardcover 1st edition 1981

Table of Contents

Preface

Mathematics education is one of the most important but least understood subjects of our age. As science and technology move the world from the age of machines to the age of computers, basic education in the language of science, technology and computers takes on increased importance. In both developed and developing nations, more people than ever before are seeking education in mathematics. Yet there are numerous signs that world-wide mathematics education is of very uneven quality, not attuned to the needs of contemporary society:

-- declining scores on standardized exams;
-- diminishing number of certified mathematics teachers;
-- public outcry at failures of the "new math";
-- professional concern with problem solving and applications of mathematics;
-- uncertainty about the relation of computers and calculators to mathematics instruction.

It was in this context of rising expectations and mounting problems that over 2000 mathematicians and mathematics teachers from around the world gathered in August, 1980, at the University of California in Berkeley, California, for the Fourth International Congress of Mathematical Education (ICME IV). Over 400 speakers from 100 different countries addressed dozens of

different issues confronting mathematics educators
around the world:

-- What should constitute universal primary education? How should primary education be structured
   to meet the needs of both those for whom it is the
   end of formal education and those for whom it is
   just a stepping stone to higher education?

-- What role should applications play in mathematics
   curricula? Is the traditional division between
   mathematics and science instruction still valid?
   How can mathematical topics be related to cultural
   issues that are relevant to students?

-- How can calculators and computers be utilized
   effectively in teaching mathematics? How much
   will computer methods change the content of
   mathematics education? Will algorithms, programming and new methods of discrete mathematics gradually replace traditional topics?

-- How does learning language influence the learning
   of mathematics? What handicaps are imposed on
   mathematics education by shifting from a home
   language to a different school language in bilingual societies?

-- Does the current curriculum provide adequate
   training in problem solving? What does problem
   solving really mean?

-- Is geometry as a school subject dead? What could
   revive it, or replace it?

     These questions are obviously of great importance
to mathematics teachers and school administrators. How
teachers and administrators deal with them will affect
the quality of education for all children; in the long
run, it will affect the level of scientific research
and technological productivity. Yet conferences on
mathematics education are typically hidden from public
view: rarely do teachers' debates over mathematics education attract the public attention that their importance warrants.

   In _Teaching Teachers, Teaching Students_ we hope to
illuminate for the general public some of the issues
discussed at ICME IV. This is not an official report
of the Congress, but a collection of personal observations by some of the participants. Our aim has been to
highlight issues that should concern the public, and to
do it in an attractive and timely manner. In this
sense the volume complements and perhaps introduces the
official Congress _Proceedings_, which will be published
in due course (by Birkhäuser Boston) with complete
texts of many of the papers presented at the Congress.
We hope that _Teaching Teachers, Teaching Students_ will
reach many teachers, administrators, and concerned
members of the public who might not otherwise know
about the Congress.

   In preparing for this volume, the editors arranged
for several roundtable discussions at ICME IV to draw
participants into vigorous debate on the major issues
of the Congress. Reports from these roundtable discussions form the basis for five special reports in the
last section of this volume. These discussions helped
frame the contents of the entire volume by bringing
into focus the many themes that naturally arise in an
International Congress of this size.

   Teaching Teachers, Teaching Students opens with
profiles of five individuals who shaped the issues facing this Congress--Henry Pollak, Chairman of the Program Committee, and the four plenary speakers, in the
order of their presentations: Hans Freudenthal, Hermina Sinclair, Seymour Papert, and Hua Loo-Keng. These
profiles are followed by special articles written by
four distinguished mathematics educators--James Fey of
Maryland, Ubiratan D'Ambrosio of Brazil, Zalman Usiskin
of Chicago, and Bienvenedo Nebres of Manila. The middle of the volume contains several special features: a
profile of George Pólya, Honorary President of the
Congress; vignettes of some of the Congress minicourses
that introduced new aspects of mathematics and its
applications; commentary from various participants on
such popular themes as problem solving, minimal competency, applications of mathematics, and the "death"
of geometry; and an historical perspective on the
International Congresses on Mathematical Education.

The volume concludes with reports from the roundtable
discussions that highlight some of the major themes of
the Congress.

    We would like to thank Anthony Barcellos for his
able and forthright assistance in helping us cover the
Congress; Gerald Alexanderson for developing the profile for George Pólya; Murray Klamkin for photographs
from the Congress; Mary Kay Peterson for typing the
volume; St. Olaf College for use of their academic computer to produce the volume; and Klaus and Alice Peters
of Birkhäuser Boston for encouragement in undertaking
this unprecedented example of mathematical journalism.
The entire volume was written, edited and produced in
the space of four months. We hope its timeliness makes
up for whatever errors or omissions remain.

                          Lynn Arthur Steen
                          St. Olaf College
                          Northfield, Minnesota

                          Donald J. Albers
                          Menlo College
                          Menlo Park, California

                          December 15, 1980

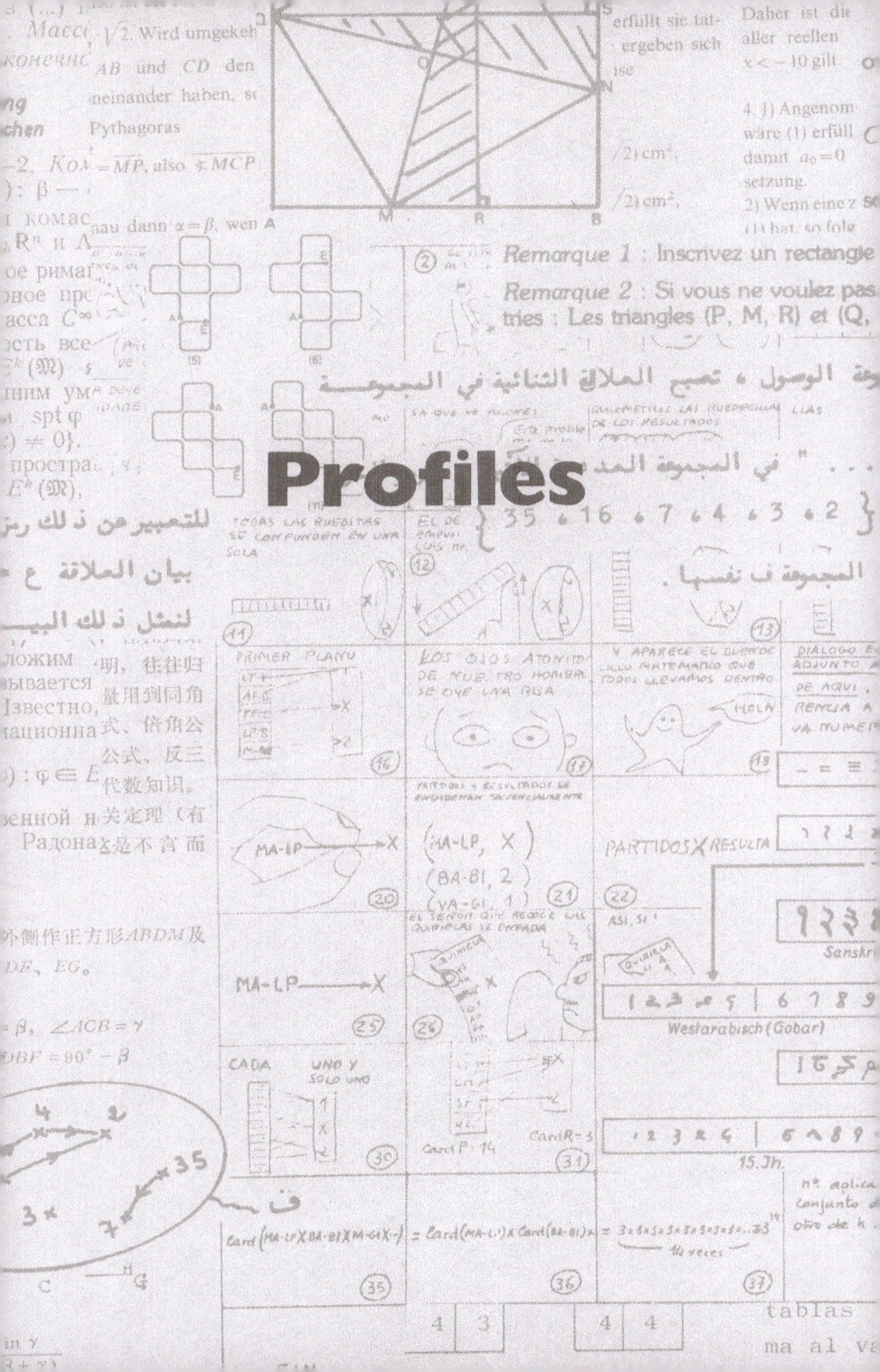

Profiles

Henry Pollak

Henry Pollak, Director of Mathematics Research at
Bell Labs in Murray Hill, New Jersey, served as Chairman of the program committee for ICME IV. Dr. Pollak
is a research mathematician who heads one of the
world's most prestigious and innovative industrial
research groups. Yet he devotes a considerable part of
his professional energy to mathematics education,
because the Bell System, like industry generally, has a
tremendous stake in education. Indeed, Bell Labs
spends each year about $1000 per employee on continuing
education. "It is not at all surprising that Bell Labs
should have a major interest in education, because it
is so important to us. My personal conviction is that
I ought to devote some time to education. I see the
result of the educational system as we hire people, and
I have ideas as what needs to be done."

"One of the great needs of our time is to rethink
curriculum and pedagogy in terms of current social
needs and technological possibilities. We should never
assume that we've got the best curriculum. At the
moment we have opportunities from calculators,
microprocessors, computers, and television in various
forms. I am not saying that the right thing to do is
to just take the technology and use it for teaching. I
am saying: Where are the pedagogical problems? What
are the current priorities? Let's see how the technology that we have available can help us. The biggest

Henry Pollak

need at the moment, in my opinion, is to have a good,
thorough look at the total elementary and secondary
curriculum in the sciences, in the social sciences, and
in mathematics, to see how the priorities of topics,
the pedagogical possibilities, and the interaction
among the topics can change in light of current technological possibilities. We have now one or two more
things to use that we didn't have before. How do they
change what we ought to do? Let's make use of them.
It's a great possibility that we've got here. I know
that it is not very much in fashion currently to support a major effort of this kind to take a global look
at curriculum and pedagogy to see how priorities ought
to change. Nevertheless, I think I would like to see
that done more than any other single thing."

        Indeed, the commitment to educational reform that
used to be very popular in the United States is now on
the defensive: popular concern over failures of the
"new math" has produced a backlash against innovation
that makes response to current problems difficult.
"One of the great discoveries of this Congress," said

Pollak, "is how much of the leadership in mathematics education is now outside the United States."

*   *   *

"Another thing I would like to see done is to make sure that college and university students turn out to be versatile people. So often there are fashions, where everybody goes into one field, so a few years later that field gets the reputation of being oversubscribed and everybody wants to go into another field. The swings in the availability of people are much too great. I would urge students to major in and to take what they enjoy. Don't let somebody talk you out of it because currently there are supposed to be no jobs. By the time you get through, it will be four or five or eight years later, depending on how far you go, and by that time the situation will be totally different. If you are a versatile person, if you haven't made up your mind that there is only one thing that is interesting, but you have kept up an interest in several, there will always be an opportunity.

"I remember that one year in the early 1970's, in the midst of a mild recession when jobs were very difficult to find, I was allowed to make more offers for people to join mathematical research at Bell Laboratories than I had in any other year. But fewer than half of those we made offers to came. Now why was that? It was because they were extraordinarily versatile people. They were people who knew both mathematics and computer science, or both mathematics and economics, or mathematics and electrical engineering very well. No matter how hard the times, such versatile people, such people who know two things and the interface between them, are going to be very much in demand. So I urge students to maintain a multidisciplinary point of view. You won't always be encouraged by your professors in that, but don't let that bother you. There will be faculty around who don't believe in computers but that doesn't mean that you have to disbelieve in computers. Maintain that breadth of interest, be prepared to be interested in more than one field, and you will do very well indeed."

Hans Freudenthal

The first plenary speaker of ICME IV was Hans
Freudenthal, former Director of IOWO, the Insitute for
Development of Mathematics Education in The Nether-
lands. A research mathematician, Freudenthal has been
a major contributor to the field of mathematics education, and it was during his presidency of the International Commission on Mathematical Instruction that the
first ICME was organized in Lyon, France. At ICME IV
Freudenthal opened the Congress with a special survey
address on "Major Problems of Mathematical Education."
Here are some excerpts from that address:

"From olden times mathematicians have posed problems for each other, both major and minor ones: here is
the problem, solve it; if you can, tell me; I will
listen to check whether it is a solution. In education, problem solving is not a discourse but an educational process. In mathematics, the problem solvers
are mathematicians. In education, problems are properly solved by the participants in the educational process, by those who educate and by those who are being
educated.

"Major problems of education are characterized by
the fact that none can properly be isolated from the
others. The best you can do at a given moment is to
focus on one of them without disregarding others."

*   *   *

"Diagnosis and prescription are terms borrowed
from medicine by educationists who pretend to emulate
medical doctors. What they do emulate is medicine of a
foregone period, which is the quackery of today. Medical diagnosis in former times aimed at stating what is

Hans Freudenthal

wrong, as do the so-called diagnostic tests in education. True diagnosis tells you why something went
wrong. The only way to know this is by observing the
child's failure and trying to understand it.

    "It is a problem, indeed, why many children do not
learn arithmetic as they should, and it is a major one
because more than anything else, failure in arithmetic
may mean failure at school and in life."

                    *   *   *

    "The history of mathematics has been a learning
process of progressive schematizing. Youngsters need
not repeat the history of mankind, but neither should
they be expected to start at the very point where the
preceding generation stopped.

    "I think that in mental arithmetic of whole
numbers we can fairly well describe schematizing as a
psychological progression, or rather as a network of
possible progressions, where each learner chooses his

own path or all are conducted the same way. The idea
that mathematical language can and should be learned in
such a way--by progressive formalizing--seems entirely
absent in the whole didactical literature."

                    *   *   *

    "There is a tendency for the learner's insight to
be superseded by the teacher's, the textbook writer's,
and finally by that of the adult mathematician. The
wrong perspective of the so-called New Math was that of
replacing the learner's by the adult mathematician's
insight.

    "It is a most natural thing that once an idea has
been learned, the learner forgets about his learning
process, once a goal has been reached, the trail is
blotted out. Skills acquired by insight are exercised
and perfected by--intentional and unintentional--
training."

                    *   *   *

    "To a large degree, mathematics is reflecting on
physical, mental and mathematical activity. The origin
of proving theorems is arguing what looks obvious.
Nobody tries to prove a thing unless he knows it is
true. This he knows by intuition, and the way to prove
it is by reflecting on his intuition."

                    *   *   *

    "Environment involves space, objects in space, and
happenings in space. The mathematicized spatial environment is geometry, the most neglected subject of
mathematics teaching today. For centuries geometry in
the English terminology was synonymous with Euclid.
But in history geometry started long before Euclid, and
in children's lives it starts even before kindergar-
ten."

                    *   *   *

    "Technology influences education. Calculators are
being used at school, and they will be used even more

in the future. Computer assisted instruction has still
a long way to go even in the few cases where it looks
feasible. ... What I seek is neither calculators and
computers as educational technology nor as technological education but as a powerful tool to arouse and
increase mathematical understanding."

Hermina Sinclair

Psychologist Hermina Sinclair was the second of
four plenary speakers featured at ICME IV. A colleague
of the late Jean Piaget at the University of Geneva in
Switzerland, Sinclair has distinguished herself in the
field of child psychology. In her plenary address she
dealt with the young child's acquisition of language
and understanding of mathematics, noting that the
development of language skills and the mastery of
mathematical ideas are more closely related than is
commonly realized.

"Children everywhere, at least in societies where
they go to school, start learning to read and write,
and to do arithmetic on paper, around the age of six.
How do children deal with this task? What previous
knowledge do they bring to it? Schools generally seem
to regard the two tasks as independent of each other;
little attention is paid to possible confusions between
the different systems that underlie spoken language,
written language, spoken numerals and computations.

"Generally it appears that educators see a big
difference between written language and written arithmetic. Children can talk before they go to school, and
it is thought that all they have to learn is to trans-
pose spoken language into written language, and vice-
versa. On the other hand, it is thought that children
cannot add, subtract, multiply or divide: this is what
they have to learn, and the notation of these

Hermina Sinclair

operations is not supposed to create any particular
difficulties. If difficulties are there, they exist at
the level of the operations themselves.

"Already at the age of 3 or 4 many children have
singled out letters (and numbers) from the other squiggles they see around them, such as decorations on wall-
paper, patterns on cloth, etc. Curiously enough, children think at first that the meaning of the squiggles
has to do with some quantitative property of what is
symbolized: when there are three dogs in a picture,
they think that there should be three squiggles in the
writing underneath it, each squiggle representing one
dog; they also may think that elephant should be written with more letters than butterfly, and that the name
of a child three years old should have three letters.
Thus, in the search for the meaning of texts, the first
link children think of is in the domain of number and
measurement, not in that of sound!"

*   *   *

"The assumption made for reading, that is, that
the child knows how to talk and merely has to learn to
put speech down on paper, has no counterpart in
mathematics. Nobody seems to think that children
already know how to add, subtract, multiply and divide
before they come to school, and that all they have to
learn is to do pencil-and-paper sums. On the contrary,
in most countries arithmetic is taught as if the conceptualization of arithmetic operations were the same
as their written symbolization. Schools do not seem to
envisage that the conceptualization of addition, sub-
traction, etc., may be a cognitive task separate from
that of writing equations, and that the latter may
present difficulties of its own.

"The number concept is constructed gradually (by
the child): the process takes many years, but it starts
very early. By only focusing on certain properties of
the number system, such as counting, schools both
overestimate and underestimate children's conceptual
capacities for dealing with numerical operations.

"Available research suggests the following conclu-
sions:
(1) Children's acquisition of number concepts and
    numerical operations (as apart from the notational
    system) is a lengthy process, stretching over many
    years, of which many details are still unknown.
(2) Just as for written language, young children have
    theories about written numerals and equations.
    These theories may not fit the symbolic nature of
    the systems elaborated by humanity over many cen-
    turies. In both domains children seem to re-
    invent or to re-construct the systems their
    societies have adopted."

                    *   *   *

"If, as I have argued, children actively recon-
struct the notational systems that symbolize spoken
language and numerical operations, we should compare
the two notational systems as to their differences and
similarities, with a view to helping children avoid
unnecessary confusions and so as to profit from progress in one system when building up the other.

"It seems to me that young children can only learn
arithmetic if they can attach meaning to numerals and
equations. Arithmetic, like reading and writing, has
to do with the extraction and construction of
meanings--at least for children. The difficulty lies
in deciding what meaning equations can have for young
children. A simple translation into words is no help.
From all we know about children as constructors of
knowledge, mathematical meanings are constructed as
action-patterns, first on real objects and later
interiorized. However, much research, and much careful
observation is still necessary on this last point."

Seymour Papert

"We are at the beginning of what is <u>the</u> decade of
mathematics education. I think the 1980's are going to
be a turning point in the development of mathematics
education on all sorts of levels. We will see dramatic
changes in what children learn; we will see subject
matters that formerly seemed inaccessible or difficult
even at college level learned by young children; we
will see changes in where learning takes place, and in
the process of learning itself. Above all, we will see
integration of mathematics into the whole person, with
an emphasis on aesthetics, on feelings, on emotions,
and on culture in a way that has not been part of our
society for a long time. The challenge to the community of mathematics educators is fantastic.

"Back in the 1960's we mathematicians and mathematics educators decided to impose something that we
thought was for the good of the people and for the good
of the children. We decided, for various reasons, that
it would be good for children to know set theory and
logic, so we decided to pass these good things on to
that culture out there. But it didn't take.

Seymour Papert

"I think the situation now is reversed. The world
out there, for a first time in a long time, really
wants something mathematical. There is a real change
out there. If we can get ourselves together enough to
listen to what is going on, I think that we can change
the relationship of mathematics to culture. And that's
what we have to do if we are to produce any radical
change in the learning of mathematics."

So began the Congress's third plenary address, a
vision of a not-so-distant future by Seymour Papert in
which computer culture becomes a carrier for mathematical knowledge. Papert is a mathematician and psychologist at the Artificial Intelligence Laboratory of the
Massachusetts Institute of Technology, where he directs
project LOGO, an innovative NSF-supported attempt to
develop a geometric language for computers that can be
learned and manipulated by very young (pre-school)
children.

Papert recounted his experience with a four-year-old named Robin who was playing computer games. One day, when her teacher came to give the computer the necessary instructions to change from one game to another, Robin said, "Let me do that." Robin's teacher was inspired enough to say OK, and wrote down on a piece of paper the necessary symbols.

"Robin couldn't type, couldn't read, couldn't write. But she began pecking away at the keyboard, copying these things she saw the teacher do. All of a sudden Robin found herself in a position where she was using alphabetic language. She was using that alphabet to make something happen, and I think in doing that she was showing us what might be a very important insight, a very important window onto the future.

"It is hard to say exactly what Robin was learning, but she was clearly learning something about the written language. And the question that this raises for me is very deep. Let's pose it like this: Why is it that children learn to speak the vocal language so early and the written language only after such a long delay? It is assumed by the educational system that this is a natural thing to be expected, but what are the reasons?

"I think the important reason is cultural. If you think of the lives that we have built for children, there is no place in this world of the child for the written language. I think that is a plausible explanation of why children don't learn to write very early. What's the point? It does not serve a purpose. For Robin it suddenly served a purpose. It gave her control, gave her power to do something that she could not do without it."

                    *   *   *

"The static image of the computer in education is as a teaching machine. Clearly what I am talking about has nothing to do with that. The most common context for computers in this conference is calculator-and-computer, or computer-and-calculator as if it were one word. It's as if computers have something to do with

number crunching, with algorithms, with all that sort
of stuff. Sure they do, but that's an infinitesimal
slice of what their real importance is, and it is very
far from those manifestations of the computer that are
likely to touch on the deepest and most fundamental
issues. For Robin the computer was not a number
cruncher, it was not a sort of super calculator, it was
not a teaching machine. It was an expressive medium, a
pipeline to an important pocket of knowledge--alpha-
betic representation of language--from which she had,
until then, been excluded."

                    *   *   *

     Papert described children playing with "turtle
geometry," part of the geometric computer language
LOGO. The children make shapes move on the computer
screen by typing simple commands with numbers to control speed and direction. Then, at a certain point,
they try using negative numbers--and the shapes all
reverse direction, collapsing to their original posi-
tions. The children's surprise blends with awe at this
unexpected power: "Negative numbers did that! Negative
numbers did something aesthetically dramatic. An idea
that was ritualistic becomes an idea of power to do
something, an idea that resonates with deep feelings,
with aesthetics, with desires."

     "As we watch these children, it becomes inconceiv-
able that anybody could be thinking about mathematics
education ten years into the future without computers
being very prominent. I mean, of course, this kind of
use of the computer out there in the world, not a
teaching machine in the classroom, but a use that
changes the kind of initial contact between the child
and the computer. These children learning to write
programs are doing formal, written mathematics at an
age when they are just beginning to make contact with
numbers."

Hua Loo-Keng

The final plenary address at ICME IV was given by
Hua Loo-Keng, Vice President of Academia Sinica and the
dominant mathematics figure in the People's Republic of
China. Hua is best known in the West for important
work in the theory of numbers--the archetype of pure
mathematics--and in the East for work in popularizing
applications of mathematics among the peasants.
Although Hua spent some years in the United States
after the end of WW II, he returned to China in order
to apply mathematics to help raise the standard of living of the Chinese people, "because I came from the
poorer class myself." In the last 20 years he has
travelled to 23 of China's 30 provinces, lecturing to
millions of factory workers and farmers on ways in
which simple mathematics can make their work more productive.

"Specialists or experts and factory workers do not
often share a common language. My experiences have
shown me that in order to achieve a common language,
these two groups must look for their common need.

"It has been my consistent aim to present the
users with the most efficient techniques. Towards this
end, my experiences have shown me that a deep understanding of the theory behind each technique is an
absolute prerequisite, lest one might be completely
misled."

Hua discussed several examples of how mathematics
can help workers solve practical problems:

   -- finding the surface area of a mountainous region
      from a contour map;
   -- finding gear ratios (within suitable mechanical

Hua Loo-Keng

   limits) to provide a speed as close as possible to
   a specified target;
-- estimating the surface area of plant leaves in
   experimental agricultural plots;
-- finding the optimum level of a process that varies
   continuously, by performing successive experiments
   at different levels, but as few of them as possible.
Each of these arose as practical problems in Chinese
factories and industries, and each requires modestly
sophisticated "pure" mathematics for its solution.

   Despite Hua's efforts to focus mathematics on
direct needs of Chinese workers, he did not fare well
during the Cultural Revolution, a period he described
as "really disastrous." "Pure mathematics was attacked
bitterly, and even applied mathematics failed to
prosper." According to Hua, The Gang of Four, leaders
of the Cultural Revolution, were interested "less in
production than in power." They in turn attacked Hua's
missionary efforts in the provinces as "sightseeing."
But Hua survived, protected by Chou En-Lai who, when

asked to describe the type of scholar that the Cultural
Revolution should encourage, responded succinctly:
"People like Hua Loo-Keng."

Hua's work started on a small scale, but the
effectiveness of his techniques quickly commanded a
large audience. "It is essential to start work on a
small scale, for example, in a workshop. If my suggestions turned out to be effective they would naturally
attract wider attention and might soon spread to other
workshops, and then to the whole factory or even to a
whole city or a whole province. In this way, my assistants and I sometimes had more than a hundred thousand
people in the audience!"

                    *   *   *

"If I were asked to say in a few words what I have
learned in these last fifteen years of popularizing
mathematical methods, I would without hesitation say
that they have enabled me to appreciate the importance
of the dictum:

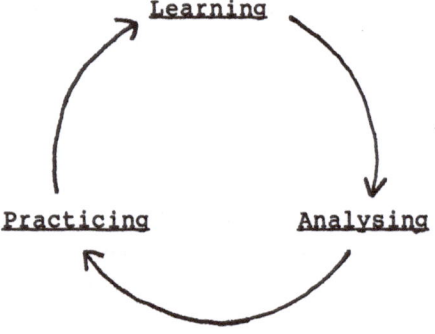

Articles

−1. Macc... √2. Wird umgekeh...
конечнс AB und CD den
lung neinander haben, s...
zeichen Pythagoras
1—2. Ko... = MP, also ∢MCP
(β): β −...
и комас... nau dann α = β, wen A
A_nR^n и Λ...
тное рима...
горное про...
класса C^∞
мость все...
E^k(M) ...
ешним ум...
ем spt φ
(x) ≠ 0}.
R) простра
a E^k(M),

Remarque 1 : Inscrivez un rectan...
Remarque 2 : Si vous ne voulez
tries : Les triangles (P, M, R) et

erfüllt sie tat-
ergeben sich
ise

Daher ist
aller reeller
x < −10 gil...

4. 1) Angeno
wäre (1) erf...
damit a_0 =
setzung.
2) Wenn ein
(1) hat so f...

/2) cm²,
/2) cm²,

Известно,
量用到同角
рациональна 式、倍角公
公式、反三
(φ) : φ ∈ E 代数知识。
твенной и 关定理（有
ме Радона 是不言而

$\angle DBF = 90° - \beta$
$= \beta, \angle ACB = \gamma$

TODAS LAS RUEDITAS
SE CONFUNDEN EN UNA
SOLA

EL DE
CADDIU
CLAS M.

35 · 16 · 7 · 4 · 3 · 2

PRIMER PLANO

LOS OJOS ATONITO
DE NUESTRO HOMBRE
SE OYE UNA RISA

Y APARECE EL CUERPO
CILLO MATEMÁTICO QUE
TODOS LLEVAMOS DENTRO

DIALO
ADJUN
DE AQ
RENTS
VA M...

MA-LP → X

(MA-LP, X)
(BA-BI, 2)
(VA-GI, 1)

PARTIDOS × RESULTA

ASI, SI.!

MA-LP → X

EL SEÑOR QUE RECOGE LAS
QUINIELAS SE ENFADA

CADA   UNO Y
SOLO UNO

Card P : 14   CardR = 3

Westarabisch (Gobar)

15. Jh.

$Card(MA\text{-}LP \times BA\text{-}BI \times M\text{-}GI \times ...) = Card(MA\text{-}LP) \times Card(BA\text{-}BI) \times ... = 3 \times 3 \times 3 \times 3 \times 3 \times 3 \times ... 3$
14 veces

Insights from ICME IV
for U.S. Mathematics Education

James T. Fey
University of Maryland

In December 1979 Izaak Wirszup, Professor of
Mathematics at the University of Chicago, submitted a
memorandum to the U.S. National Science Foundation out-
lining recent changes in science and mathematics education programs of the Soviet Union. Based on his extensive studies of Soviet technical and educational
literature, Wirszup concluded that graduates of Soviet
secondary schools now receive so much more mathematical
and scientific training than U.S. students that the gap
poses a formidable challenge to American national secu-
rity. Wirszup's report soon circulated widely; it was
noted by science journalists, and led to major news
stories in many cities. These ominous reports of
foreign progress in education exacerbated the continuing public criticism of U.S. education, re-awakened
memories of earlier studies showing superb performance
of Japanese mathematics students but only modest
achievement of U.S. young people, and sent many American mathematics educators to the Fourth International
Congress on Mathematics Education with a mixture of
curiosity and concern about our national effort in
mathematics teaching.

The program for ICME IV was rich with potential,
promising presentations by acknowledged world experts
in mathematics and its teaching at all levels of
schooling. The Congress provided many sessions on
geometry, algebra, calculus, applications, computers,

research, statistics, problem solving, and teacher education. The Berkeley venue was attractive, the hospi-
tality provided by the local organizers was superb, and
the openness of all Congress participants made possible
easy and lively communication across cultural and
linguistic barriers. For each participant in the
Congress, the experience was undoubtedly a personally
pleasant and stimulating professional event. But, in
retrospect, the intellectual substance of the Congress
was disappointing for mathematics education. If some
country holds the key to effective, exciting school
mathematics, the secret was not revealed at ICME IV.
Instead, the prevailing mood was anxiety about the
problems faced by mathematics teachers; the approaches
to these problems were most often tentative and con-
servative. Mathematics education seems to be in a
slump paralleling much of the world's economy.

Some of the disappointment in this International
Congress was undoubtedly due to the absence of Soviet
mathematics educators and very limited contributions
from Japan, China, and Eastern Europe--all countries
from which one would expect different cultural traditions to produce different approaches to schooling.
Nevertheless, the past 20 years have produced exciting
innovations in the Americas, Africa, and Western
Europe, so the Congress still held promise of reports
on major new developments.

If Misery Loves Company

One of the perverse pleasures in talking to
mathematics teachers in other schools or other countries is the realization that, bad as things might seem
to be here, problems are worse elsewhere. The Congress
program was planned by an international committee, and
it included well-attended talks and panel discussions
on such familiar American problem subjects as minimal
competencies, decline in numbers of undergraduate
mathematics majors, successes and failures of recent
curricula, participation of women in mathematics, use
and misuse of textbooks, roots of failure in primary
school arithmetic, the death of geometry, and changes
in mathematical preparation of students entering post-
secondary schooling. While the problems often seem

similar, the causes and professional responses vary
from country to country in ways that are illuminating.

Just prior to the last ICME in Karlsruhe, West
Germany, it became evident that the issue of minimal
mathematical competence was of concern in several European countries, in addition to the United States. A
study group formed at that meeting led to a sharing of
information that revealed deep concern about the challenge to schools in several dozen countries. However,
the problem had quite different foci in different
nations and professional reaction to the call for
guarantees of minimal competencies took on decidedly
different characteristics in different countries. In
many European countries, social pressure to democratize
educational opportunity has greatly expanded the number
of students continuing into comprehensive (as opposed
to vocational) secondary schools; in several newly
independent developing countries of Africa, similar
political pressures have expanded the pool of secondary
students. In both situations there is concommitant
societal concern that the academic standards of a pre-
viously elite system are being compromised in the move
toward democratization. This concern is often
expressed as a call for some guarantee of minimal
achievement standards (usually far beyond those of
interest in the U.S.).

For American mathematics educators, this problem
--new to Europeans and Africans--is a very familiar
one: how to cope with a very diverse school clientele,
including less able and more reluctant secondary school
mathematics students. Other countries have not found
magic solutions to the difficulties, although in Scot-
land and England the concern about minimal competencies
has prompted broad consultation between educators and
employers. There is little precedent for such consultation in the U.S. and we might profit from careful
study of the process and products of such efforts.

What is particularly striking about the minimal
competence movements in other countries is their
approach to assessing student competence. In the U.S.
we have almost uniformly relied on objective paper and
pencil tests to measure attainment of competencies.

Further, these tests are accompanied by the thinly
veiled implication that teachers and systems whose students do not reach suitable test score levels are not
doing their job. The European attitude toward measurement of competence is strikingly different. While
school-leaving exams are important in many countries,
it seems clear that the judgment of teachers is still
given very substantial weight in appraisal of student
achievement, and the idea that standardized multiple-
choice exams can accurately evaluate effectiveness of a
school program has made very little headway.

Problem Solving vs. Standardized Testing

There can be little doubt that increased use of
standardized testing in the United States has given us
a much clearer national picture of achievement patterns. It is also possible that countries which do not
monitor student progress so closely are only unaware of
bad news. However, evidence from comparative international studies of mathematics achievement generally
show good results in Japan, Germany, Belgium, and
Israel, for example. This leads one to skepticism
about the potential benefits of extensive U.S. testing
programs. It also suggests a number of conjectures
about the pernicious effects of emphasis on testing and
accountability. First, typical examination questions
used in European schools are not short-answer or
multiple-choice. Instead, they call for multi-step
problem solving and analytical work, plus presentation
of the results in clear written form. Recent National
Assessment results in the United States suggest that
American young people are particularly weak in problem
solving. Has our reliance on objective testing slanted
patterns of school performance away from goals that
would get almost universal endorsement by mathematics
teachers?

In many school systems the implementation of
standardized achievement monitoring has been designed
to guarantee a more uniform and coherent school curriculum. There is ample evidence to suggest that the
actual result has been, in many cases, teaching to the
tests. While the virtues or sins of this practice are
debatable, it creates another largely unnoticed effect

on schools. The best and most creative teachers seldom
find any pleasure in a situation that is heavily con-
strained by externally prescribed syllabi and examinations. Recent NSF status surveys of school mathematics
and science confirm the apparent impact of such constraints (see [2] for summaries of these surveys).
There was evidence of a growing militancy resisting
accountability, evaluation, and bureaucratization of
schools. One teacher spoke sharply,

> We need to be working with teachers, not
> checking on them. Education is generally a
> negative enterprise toward children, toward
> teachers. It is a highly structured reward
> structure which emphasizes the negative.
> Those who get rewarded are those who make the
> fewest mistakes. [1; p. 8]

These remarks might exaggerate attitudes of American mathematics teachers, but conversations at ICME IV
strongly suggested that in many other countries the
professional judgment and autonomy of mathematics
teachers is more highly regarded and respected. A
French mathematics teacher said, "Le professeur est le
roi." ("The professor is king.") Others reported that
teachers in their countries simply would not submit to
heavy-handed accountability measures.

While this comparison of teacher professional sta-
ture and self-concept might well overstate the disen-
chantment of American teachers and the independence of
those in Europe, it is interesting that the phenomenon
of teacher burnout and disillusionment was not mentioned prominently except by Americans. Just as we
acknowledge that a learner's self-esteem and self-
concept strongly influence their performance in school,
it seems that we have ignored that factor in the professional lives of teachers.

Social Context of School Mathematics

The existence of an International Commission on
Mathematics Instruction (ICMI) and the quadrennial
Congresses that began in 1969 are testimony to the
emergence, over the past 25 years, of a large number of

people whose professional specialty is mathematics education. By training and inclination it is natural for these specialists to look within the discipline of mathematics and the mathematics classroom for ways to improve teaching of mathematics. But trends like teacher burnout or declining student achievement and motivation are not limited to mathematics. They suggest that effectiveness of any one special instructional program is affected by a complex of broad school and societal factors. At ICME III in 1976, Douglas Quadling of England expressed the hope that the program of ICME IV would help mathematics educators see ourselves as others see us. There were in fact numerous sessions at ICME IV devoted to applications of mathematics, including a few presented by representatives of various sciences. A major plenary session talk by Hermina Sinclair outlined connections between the development of language and mathematics, and another by Hua Loo-Keng described Chinese experiences in popularizing mathematical methods in various applied fields.

The construction of bridges between school mathematics and the sciences in which it plays such a central role has long been a topic of interest in mathematics education. The formal program of ICME IV gave little attention to mathematics in broader school and social contexts, but several talks alluded to the importance of these factors. For instance, reporting on mathematics at the college level in Southeast Asia, Bienvenido Nebres remarked that in countries where mathematics is flourishing there is a long-standing cultural tradition of regard for intellectual activity; in countries struggling to develop high level mathematics, that tradition for intellectual accomplishment is absent. Informal conversations with mathematics educators from Europe confirmed the important role of such societal regard for academic achievement, indicating, for instance, that in Europe undergraduate choices of major fields seemed less sensitive to immediate employment prospects than in the United States (although even there computer science/informatics is clearly attracting many would-be mathematics majors).

Although recent survey data suggest that Americans place high value on school mathematics achievement for

their children, there are strong suspicions that this
attitude reflects a limited view of mathematics (arithmetic for practical affairs) and a vague awareness that
mathematics is a prerequisite for entry into many
rewarding occupations. Teachers and administrators
perceive a decline in community and parental support of
the schools and recognize clearly the debilitating
impact that family and social instability has on school
programs. For example, in a recent monograph analyzing
mathematics education in the Soviet Union, Robert Davis
commented extensively on the markedly different
school/family relations in that country. He noted that
Soviet families tend to watch school progress of their
children very closely and that school/home relations
are active and close [3]. These international con-
trasts suggest that it would be productive for mathematics educators to pay more attention to their role in
the total school program and for schools to initiate a
searching dialogue with the communities they serve cov-
ering goals and patterns of schooling--not simply bus
schedules and tax levies. We could probably learn a
great deal from recent experiences in England, Germany,
and France of such broad public study of educational
policy.

Diamonds in the Rough

In a Congress dominated by statements of problems for which only modest and hesitant solutions were
offered, there were a number of bold, exciting presentations. One of the best was certainly Douglas
Hofstadter's "Analogies and Metaphors to Explain
Gödel's Theorem." The Pulitzer Prize winning author of
Gödel, Escher, Bach presented a marvelous collection of
approaches to understanding one of the most disquieting
findings of formal mathematics: any axiomatic system
adequate for derivation of familiar mathematics contains statements whose truth or falsity cannot be
decided. Hofstadter illuminated the central issue,
self-referent statements, by a series of familiar
illustrations,

$$\boxed{\text{T H I M K}}$$

$$\boxed{\text{P L A N \quad A H E AD}}$$

and an original:

> **Hofstadter's Law**: It always takes longer than
> you think, even if you take Hofstadter's Law
> into account.

At a panel discussion on the frequently reported
death of geometry, Jean Dieudonné gave a perceptive and
comprehensive overview of the subject and its connections to advanced mathematics. In a clever dissent
from Dieudonné's Bourbakist view of mathematics, Robert
Osserman offered a 'proof' that geometry is dead:

> We know from Plato that God ever geometrizes,
> We know from Nietzsche that God is dead,
> Therefore geometry is dead.

Osserman went on to give a convincing argument that
geometry is not dead at all and a thoughtful analysis
of structuralist epistemology in mathematics.

The prospective influence of computers on mathematics at all levels of schooling was naturally a topic
of many Congress sessions. In provocative talks, Seymour Papert and Arthur Engel proposed bold visions of
how technology might remodel conventional pedagogy and
curricula. J.H. van Lint provided a beautifully lucid
exposition of algebraic coding theory, a field largely
stimulated by possibilities of modern computing
machinery.

Ivory Tower vs. Classroom Reality

These, and a few similar talks at the Congress,
represent the best spirit of what such an international
meeting ought to be about--experts in their fields making deep analyses of important problems or explaining
the latest scientific developments. However, these
same presentations raise a fundamental problem of the
field. To most teachers or supervisors responsible for
school mathematics, these bold thinkers seem to be
talking about a world that is disjoint from 99% of real
schools or real pupils. It would be very easy for
classroom teachers to listen politely and dismiss the

proposals as hopelessly naive, far removed from the
main business of elementary and secondary education.
This observation about presentations at the Congress
can probably be generalized to a broader segment of the
program, and it would be easy to dismiss the whole
enterprise as irrelevant. Instead, this gulf between
the International Congress and practical school people
is evidence of the unfortunate estrangement that typi-
fies relations between the several groups concerned
about the health of school mathematics instruction.

One of the most exciting aspects of the "new math"
era was interaction among university mathematicians,
teacher educators, school level supervisors, and classroom teachers. The interaction was not always harmoni-
ous, but when sparks flew they indicated new ideas
being born and conventional wisdom being challenged.
The products of this unprecedented collaboration were
not uniformly successful in school practice and the
energy to continue improvement was not always sus-
tained. As a result, school people often feel victim-
ized by poor advice from university mathematicians
while the mathematicians plead that their ideas were
never well understood or properly implemented.

Today university mathematicians frequently com-
plain, in imperious tones, about the dismal products of
secondary schools (criticism which secondary teachers
also direct to their elementary counterparts). If
these mathematicians offer to pitch in and work on the
problems, they are, understandably, held at arm's
length by the schools. This is a tragedy. It is
entirely possible that the ideas of Papert or Engel
indicate the wrong directions for computers to influence school mathematics; Gödel's Theorem probably has
little significance in the near future for teaching
arithmetic or algebra in high school. But Papert,
Engel, Hofstadter, van Lint, Dieudonné, Osserman, and
others are exploring new ideas, and they bring fresh
perspectives to consideration of educational problems.
Surely even the most pragmatically oriented can be
stimulated by interaction with these creative minds.

In searching for reasons why many teachers seem to
have lost their enthusiasm and spirit of innovation,

Mary Lee Smith has suggested that a teacher's position is really very isolated. As the sole adult in a class of young people it is often hard for teachers to maintain contact with the scientific and professional communities beyond the school. She noted that some had "kept a window on the larger world of ideas," but that "most teachers have only a mirror that reflects the values and ideas already dominant in the public schools." [4; p. 18] This is surely an undesirable state of affairs, for any teacher at any level. The redevelopment of working relations among the various parties with interest in school mathematics should be high on the priority list of the profession.

Structure of Mathematics Education

In 1900, at the International Congress of Mathematicians in Paris, David Hilbert posed a list of 23 problems that he felt to be of central importance in mathematics. Many of those problems have now been solved. Some have not, but there is no question that the list stimulated enormously important research in the discipline of mathematics. When research mathematicians visit meetings of mathematics educators they frequently come away puzzled about the focus of the field, wondering if there are any well-defined problems or structures of established knowledge.

It is certainly fair to say that most mathematics teaching today proceeds under guidance of intuition and experiences of individual professionals, not from a coherent empirically validated theory of learning or instruction. It is also true that, until very recently, research in mathematics education has consisted of uncoordinated small-scale studies, most testing personal hunches of individual investigators. When E.G. Begle compiled a summary of research-based knowledge in mathematics education, he was able to group studies into a number of very broad categories, but within each category he offered no more cogent organization than alphabetical order by sub-topic [5]. A skeptical view of ICME IV could reasonably conclude that lack of a clear professional agenda and purpose still plagues the field. The advice given in most talks was largely opinion, buttressed by personal

experiences and observations. In the research sessions
there was a great deal of question-posing and much less
result-reporting. Not a few sessions described ideas
that have been rediscovered every 20 years for quite
some time. However, a more sanguine observer of the
Congress and of mathematics education internationally
can see some very encouraging signs of progress.
Piaget-inspired investigations of young people's intellectual development has immeasurably increased our
understanding of the processes in learning basic
mathematical ideas. Curricular resource materials
available to mathematics teachers today are far richer
than at any previous time. Educational policy makers
can be guided by empirical studies of current practice
never before available.

    Mathematics education differs in fundamental ways
from physical science; it is a problem field much more
like the social sciences, trying to understand an
enterprise in which the crucial parameters are seldom
replicable from time to time or place to place. The
International Congress did address important questions
of learning, instruction, curriculum, and evaluation.
If there is a discouraging and unyielding problem facing the field, it is the difficulty of translating best
available knowledge into common classroom practice.
For example, Max Bell assembled for the Congress a mas-
terful survey of materials available for teaching
applications of mathematics. He found the supply
impressive, much having existed for 10-15 years, but
concluded that those good ideas have made very little
progress in everyday school curricula. Instead there
is a world-wide back-to-basics movement that seeks
solutions to current problems in the practices of much
earlier days.

    In reaction to the ICME III meeting in 1976, Zal-
man Usiskin pointed out that those condemning recent
innovations and calling for return to older approaches
forget that the revolution in school mathematics of the
1960's was provoked by the fundamental inadequacies of
earlier programs [6]. There are exciting prospects and
challenges ahead in mathematics education, but to realize the opportunities and meet the challenges the profession must break out of its pessimistic, hesitant,

critical mood into bold, creative, and cooperative ven-
tures. One can only hope that ICME V in Adelaide will
be sparked by such renewed vigor.

References

[1]   Denny, Terry. "River Acres." In Robert E. Stake
       and Jack Easley (Eds.) Case Studies in Science
       Education. University of Illinois, 1978.
[2]   Fey, James T. "Mathematics teaching today: per-
       spectives from three national surveys." The
       Mathematics Teacher 72 (1979) 490-504.
[3]   Davis, Robert B.; Romberg, Thomas; Kantowski, Mary
       Grace; Rachlin, Sid. An Analysis of Mathematics
       Education in the Soviet Union. Columbus: ERIC
       Center for Science and Mathematics Education,
       1980.
[4]   Smith, Mary Lee. "Fall River." In Robert E.
       Stake and Jack Easley (Eds.) Case Studies in Sci-
       ence Education. University of Illinois, 1978.
[5]   Begle, E.G. Critical Variables in Mathematics
       Education. Washington: Mathematical Association
       of America, 1979.
[6]   Usiskin, Zalman. "ICME III: What Was It?" The
       Mathematics Teacher 70 (1977) 436-441.

Uniting Reality and Action:
A Holistic Approach to Mathematics Education

Ubiratan D'Ambrosio
University of Campinas
President, Inter-American Committee on Mathematical Education

> In the most recent upheaval [May, 1968],
> the intellectual discovered that the
> masses no longer need him to gain know-
> ledge: they know perfectly well, without
> illusion; they know far better than he
> and they are certainly capable of
> expressing themselves.
>
> -- Michel Foucault, 1972

Mathematics, the way it is practiced in school
systems today, seems destined to disappear, a casualty
of rapidly changing patterns of daily life. Already a
new educational mood reflects the impact of new technologies and the resulting changes in family structure.
This new mood was clearly evident at ICME IV.

The new technologies affect educational systems
primarily by increasing the amount of information
available through the media. This occurs not only in
those media derived from advances in electronics, but
also in the more traditional forms such as magazines,
newspapers, and cinema. Every day these become more
accessible, more attractive, more visual, and easier to
digest. The most pronounced effect of this increased
flow of information is the instantaneous worldwide communication which brings into our homes events from all

over the world in the same instant as they are happening. The widespread use of satellite communication
gives us an unprecedented global view of events. The
critical analysis of these events is done, simultane-
ously and independently, by rich and poor, by adults
and children, by teachers and learners. This leads in
a natural way to an increasing call for dialogue, for
common search for meaning, for global analysis of
events. These events are surely reflected in every
phase of the relationship between teacher and learner,
upon which the entire educational structure relies.

Information-Communication-Dialogue

Our daily life depends on a troika of "information-communication-dialogue" which deeply affects family life and, consequently, school life. The way it
affects those fundamental cells of our society is manifested mainly in a new relationship between generations, which amounts to a different concept of "giver-
receiver" in the school system.

The traditional role of the teacher who knows as
opposed to the student who does not know and goes to
school to learn is thus challenged. Both have seen the
same events; both have witnessed the same facts; both
are aware of current knowledge, in many cases to the
advantage of the student, who is often better informed
than the teacher. Both student and teacher proceed,
consciously or unconsciously, to the analysis of
events, facts and knowledge. Again the student has an
advantage: the teacher is compromised by modes of
thought which are structured by his specialty as
expressed in the traditional disciplines of the curriculum, and which allow only for a narrow and limited
analysis of a global phenomenon, such as those resulting from or composing current events.

To make up for this disadvantage, the teacher
resorts to authority deriving from his knowledge, rein-
forcing the traditional power structure in the classroom and thus moving against the growing need for
dialogue which the relationship between generations is
calling for. As a result, the classroom becomes a
painful experience for all concerned.

Information-communication-dialogue affects not
only individual relationships, but society as a whole.
Increasingly questions arise concerning the relevance
of education, costs versus benefits, and the priority
for schooling in societies with more immediate needs.
In developing societies, where information and communication are the forerunners of development, this ques-
tioning of the value of a traditional education has
become more and more common. Although at first schools
are regarded as a sure road to both personal and societal improvement, the results of schooling are slow, of
very remote benefit, and hence questionable. This
attitude was beautifully expressed by G. Ladda in his
autobiographical novel Padre, Padrone.

Summing up, I see an unpromising future for traditional formal education in this rapidly changing world.
Particularly with respect to mathematics, I sense an
unsustainable situation. But ICME IV showed some directions that may reinstate mathematical modes of thought,
hence mathematics itself, as a central subject in a
renewed school structure.

Holistic Education

We can benefit much by observing what is going
on in the developing world. In most developed countries school systems are already fully established,
attending the needs of a somewhat stable population.
In these countries evolution of the educational system
means basically improving what already exists. In a
sense, the objective is to run an on-going operation,
albeit in a better way. Hence change cannot be pro-
found. The existing structure--mainly the teacher
power structure in the classroom--is established and
very difficult to change.

In developing countries, the educational system is
a system in the making. The risk of building up a system which may be obsolete at its birth invites a deeper
analysis of what to expect from an educational system
as a whole. The problems of a global nature which per-
meate the process of development require comprehensive
planning for education involving all of a nation's
social structures: its people, its societies and its

cultures. This planning must establish future priorities in terms of current needs. This is _holistic_ education, a thorough response to the needs of growing educational systems that is absolutely essential if mathematics is to acquire a prominent role in the educational system of the developing countries.

Reality: The Starting Point

The conceptual framework in which we place ourselves in this holistic approach calls for the recogni-
tion of reality as the starting point of any educational experience. Reality for a child is the natural
and social environment in which he or she lives, an
ambience with both cultural and emotional overtones.
Reality is _perceived_ through the senses (understood in
the most general and naive way). In the same way as we
perceive reality, we perceive knowledge, we perceive
language, and we perceive arithmetic. This process was
very well described by Hermina Sinclair in her plenary
address at ICME IV.

Elementary school children frequently exhibit
rather sophisticated perceived skills in mathematics.
For example, an observational experiment conducted during a regular class for 6 year olds in Passo Fundo,
Brazil, showed that while the teacher was struggling to
teach how to add, for example, 3 + 5, resorting to
small ducks drawn on the blackboard, a group of boys in
the corner of the room were playing with small pictures
of soccer stars and performing operations as sophisticated as "Yesterday I gave you 10 pictures, now you
gave me 7, so you still owe me 3."

The same kind of behaviour was noticed during an
experiment conducted in Campinas, Brazil, to observe
reactions of 7-year-old children of lower middle class
families when exposed for the first time to hand calculators. Not only was the ability to handle the calculators immediately _perceived_ without need of instruction, but also the children revealed arithmetical knowledge far beyond that which was being taught in the
school. Surprisingly, the children showed abilities in
performing operations which, when presented in the regular formal class, led to failures in examinations.

I am sure that many similar examples of perceived
mathematics could be given. They are even more common
in the areas of language. It was stressed numerous
times at ICME IV, not only by Hermina Sinclair but also
by panel members on the session "Teaching Mathematics
in a Second Language," that perceived language often
conflicts with received language. In bilingual societies these two languages carry different structures and
different emotional connotations (such as, for example,
those resulting from the colonial situation). These
conflicts generate barriers to learning which may cause
permanent damage to the student.

As a specific situation, it was pointed out by one
of the panel members that while English and most Latin
languages use construction such as "three plus two is
five," there are languages in which the perceived knowledge, that is the knowledge derived from everyday
practice, will lead to an arithmetic in which one would
say "three will give you five when you add two." This
kind of language structure would require a different
way of teaching arithmetic.

The core objective of the educational process is
to develop the capability of critical reflection on
perceived reality and to generate strategies for meaningful action based on this reality. John Dewey's
objective of acquiring knowledge and skills can no
longer be the focal point. In holistic education knowledge and skills arise as subproducts in the process of
reflecting and acting on a perceived reality, not on a
"specially arranged" reality. It is for this reason
that holistic education is incompatible with so-called
"back to basics" movement, for in this movement the
individual's capability of perceiving reality is subject to practices which precondition him to an artificial reality.

This point is well illustrated by the situations
observed in Brazil which we referred to above.
Besides, we have to realize that learning school mathematics represents a step towards abstract reasoning
rather than abstract reasoning itself. The barriers
created against accepting mathematics in an artificial
problem situation, which is provided as motivation for

this abstract reasoning, are of the same nature as
those identified in the teaching of a foreign language
without creating the proper cultural environment.

In his critical study of scientific knowledge,
Paul Feyerabend employs "history" in the same sense as
we employ "reality":

> Scientific education as we know it today
> ...simplifies "science" by simplifying its
> participants. First, a domain of research is
> defined. The domain is separated from the
> rest of history (physics, for example, is
> separated from metaphysics and from theology)
> and given a "logic" of its own. A thorough
> training in such a "logic" then conditions
> those working in the domain; it makes their
> actions more uniform and it freezes large
> parts of the historical process as well. [1]

This describes an escape from the isolation of disci-
plinary thought that is one of the most important
directions mathematics is taking. Certainly it has to
be incorporated in schooling through more dynamic
classroom arrangements and curriculum concepts (see in
this respect [2] and [3]). Thus the holistic viewpoint
carries even further the approach of Imre Lakatos [8]
which views mathematics education as an exercise in
critical reflection upon perceived reality.

Action: The Focal Point

The holistic approach, not basically different
from a theoretical framework for modelling, can be
illustrated by the schematization in Figure 1. (For
details, in particular from the viewpoint of models,
see [3] and [4].) The holistic approach to education
requires dynamics that bring accumulated knowledge,
skills and abilities to bear on desired action. This
objective goes a step beyond what is usually called
learning by activities. It calls upon man's inherent
will to influence, to adapt and to change reality in
solidarity with his will to understand reality. The
uniting of these two inherent wills, so characteristic
of our species, is the challenge for education.

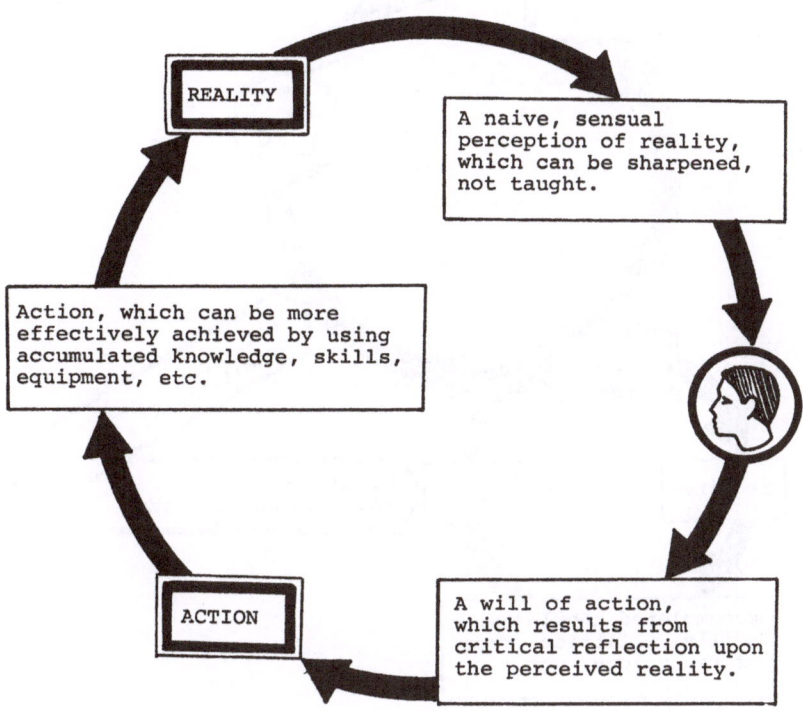

REALITY

A naive, sensual
perception of reality,
which can be sharpened,
not taught.

Action, which can be more
effectively achieved by using
accumulated knowledge, skills,
equipment, etc.

ACTION

A will of action,
which results from
critical reflection upon
the perceived reality.

Figure 1

The totality of available knowledge, technologies,
and skills is a result of a cumulative process which
spans the entire history of mankind. These accumulated
bits of knowledge, techniques, and skills are per se
dead pieces, meaningless; they become meaningful only
when placed within a context. Unfortunately, our educational practices are loaded with the mere acquisition
of knowledge and skills. Far too often these acquisi-
tions are goals in themselves.

A New Role of the Teacher

Teachers and schools must play a new role in
holistic education, a combined utilization of techniques of group dynamics and information retrieval.
The teacher must intervene in the process of learning
by stimulating critical reflection by the student upon

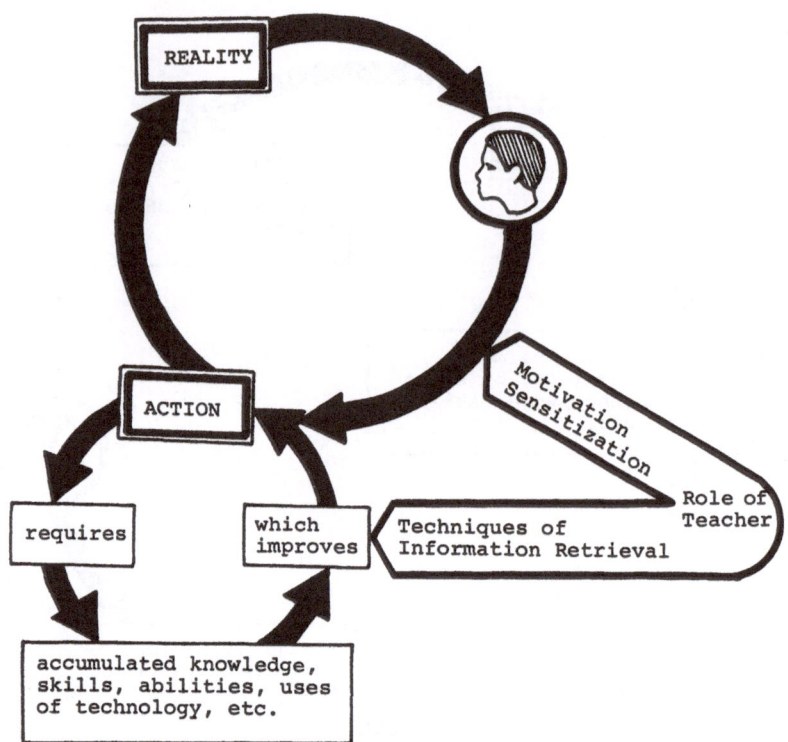

<u>Figure 2</u>

his perceived reality, which as a result will sharpen
his capability to perceive reality. Of course the
teacher would also act as an instructor in a more traditional sense, in teaching techniques of retrieving
information, techniques which include much of what
might be called basic knowledge and skills (see Figure
2). Basic skills are never introduced only for their
own sake; they enter the classroom as an instrument for
the retrieval of needed information that will make possible desired action.

The training of teachers requires, therefore, a
substantial change. It is not the teacher's knowledge
or skills that will determine his effectiveness, but
rather his ability to sense children's perception of
reality and children's drive toward action. Of course,
mathematics does not escape from this scene. Either

mathematics fits into the context in which the individual functions, or mathematics will be an appendix to the educational experience, rather a "useless appendix," using the phrase popularized by H. B. Griffiths and A. G. Howson.

A teacher's ability to stimulate the learner's intellectual drive will be matched by the learner's capability to help satisfy this drive, thus providing access to needed techniques, skills, and knowledge. But unless the learner fulfills his drive for action, he will discard schooling, teacher, and subject.

There are several methodologies which can be adapted to suit this holistic approach to mathematics education. Many examples can be developed merely by common sense, once this holistic concept is accepted. (See [4], [5], [6], and [7] for some specific examples.) Modelling is probably the approach best suited for this context, as are projects. It all depends on the teacher's attitude.

The role of the teacher is not to transfer knowledge, nor to command an action, but to show and explain reality. The teacher should be a companion in the perception of reality, and a supplier of techniques to retrieve desired information. In this approach the teacher's power as manager of the educational experience is replaced by his participation in a joint intellectual venture; his relation with the student is not based on authority, but rather on partnership in the pursuit of understanding and change.

References

[1] Feyerabend, Paul. Against Method. Verso, London, 1978.

[2] D'Ambrosio, Ubiratan. "Mathematics and society: Some historical considerations and pedagogical implications." Int. J. Math. Educ. Sci. Tech. (forthcoming).

[3] -----. "The relationship of integrated science to other subjects in the curriculum." New Trends in Integrated Science Teaching, Volume V, Chapter 4.

UNESCO, Paris, 1979, pp. 27-34.

[4]   -----. "Modelos matemáticos do mundo real."
       ["Mathematical models of the real world."] Ciencia
       Interamericana, (OAS--Washington) 20:1-2 (1980)
       4-7.

[5]   -----. "Issues arising in the use of hand-
       calculators in schools." Int. J. Math. Educ. Sci.
       Technol. 9 (1978) 383-388.

[6]   -----. "Strategies for a closer relationship of
       mathematics with the other sciences in education."
       Cooperation between Science Teachers and Mathemat-
       ics Teachers, Materialen und Studien Band 16,
       Institut für Didaktik der Mathematik, Bielefeld,
       1979, pp. 311-330.

[7]   -----. "La educación científica en el contexto
       socio-cultural." ["Science education in the
       socio-cultural context."] Working paper, Seminar
       on Integrated Science Education in Latin America,
       Huaraz, March 1980, UNESCO--Santiago, 41 pages
       (mimeo).

[8]   Lakatos, Imre. Proofs and Refutations: The Logic
       of Mathematical Discovery. Cambridge University
       Press, 1976.

Decision-Making in Mathematics Education

Zalman Usiskin
University of Chicago

At ICME IV, we were reminded that in the middle ages the long division algorithm was so new and advanced that at least one school decided to teach long division and advertised this in the hope of attracting students. This underscores two longstanding properties of the school curriculum: it is changeable and those who have made decisions to change it have done so for a variety of rationales regarding what is best for the people for whom the curriculum is intended. Yet mathematics students throughout the world seldom think of their school experiences as being shaped by people and subject to decision-making. They are more likely to believe that arithmetic and other aspects of mathematics are facts of life that have been part of the school curriculum since the beginning of time.

A congress on mathematics education of world-wide scope enables curriculum decision-makers to compare perceptions, to make use of the work of others with whom they seldom come into contact, and (one would hope) to make wiser decisions regarding the mathematics to be taught their students. Merely bringing people together serves useful purposes; "we have much to gain from sharing ideas with others" is stale but still accurate rhetoric.

However, decision-making requires more focussed direction than that given by happenstance meetings. At

a minimum, the knowledgeable decision maker needs to be
aware of issues and various viewpoints concerning them,
to distinguish issues from areas of universal agreement; needs to be able to examine these issues from a
variety of perspectives; and should have specific
information about the impacts that previous related
decisions have had on other people, other institutions,
and other societies.

Raising Issues

    One of the factors distinguishing formal educa-
tion from happenstance learning is the attention given
to the medium of instruction. In principle, one should
employ that medium most appropriate to the instructor,
to the content, and to the audience. Sadly, the
eminent educators who speak at international congresses
unwittingly provide highly visible examples of what
should not be done in education, for the medium of
instruction is frequently ineffective.

    If we are to honor our great authors, we would be
better served by asking them to write. Similarly, we
should invite to give talks only those who are noted
for their speaking. And for those who have been great
leaders but do not excel in either the written or oral
mode, it would be appropriate to have a session at
which their accomplishments are detailed and in which
they can listen to the accolades of their peers. Let
us not continue to disgrace ourselves and degrade those
whom we wish to honor by asking them to operate in
modes for which they are ill-equipped.

    From the standpoint of decision-making, plenary
session speakers should either raise or discuss particular issues of wide importance. Although Freudenthal,
Sinclair and Hua discussed areas of concern at ICME IV,
only Papert spoke to an issue. It was interesting to
note the response of those in attendance. In trying to
convince us of the power of the computer, Papert suggested that the "under-developed" countries might chea-
ply leap-frog over the "developed" countries in mathematics education by purchasing computers for all of
their citizens, a suggestion greeted with chuckles
(which, in the climate of such sessions, is a strong

response). Few in attendance at his talk had also
attended the Snowmass Conference in 1973, at a time
when four-function hand-held calculators cost around
$150. Papert suggested then that each child in the
United States be given a calculator, reasoning that the
massive quantities of calculators so produced would
lower their individual price to $5 or $10 so that there
would be more impact per dollar on students' mathematics education than would occur as a result of any
curriculum projects [1]. The reaction to this suggestion was laughter from almost all of the approximately
fifty mathematics education leaders present. Today we
would not laugh at such a suggestion and I believe the
developing countries should take his 1980 thoughts
seriously.

An issue requires choices to be made. Papert
focussed on an interesting issue: Should we provide
calculators to those who do not have them rather than
spend time teaching arithmetic? Those who happened to
also attend the session where Marilyn Suydam gave a
summary of the usage of calculators in classrooms
worldwide [2] were treated to the best rationale for a
congress like this: raising or focussing upon an important issue and having broad-based information to help
in making decisions about that issue.

Applications in the mathematics curriculum
received strong attention in several worthwhile sessions. My colleague Max Bell gave a most informative
presentation and handout at his session devoted to the
results of a worldwide survey on materials for teaching
applications at school level. It was a model for what
ought to be in a plenary session. Arthur Engel gave
three delightful talks on algorithms. Speakers on
mathematical topics, many of them recruited by Henry
Pollak from past and present Bell Labs workers, were of
uniformly high quality.

Raising an issue requires only that one side of
the issue be presented, but fruitful discussion of an
issue requires more points of view. Although I would
guess that there were some exceptions not known to me,
no one decried calculators, spoke against the omnipres-
ence of computers, or argued that statistics is not for

everyone. Despite the current popularity of Piaget's
rather pessimistic views regarding what children at
given ages can learn, no one spoke against Papert's
quite optimistic views. Almost every speaker took
pains to be conforming and polite. As a result, necessary and useful information, confrontation and discussion about some of the most visible issues in mathematics education today never surfaced.

An Issue in Content Selection: What Can Be Cut?

There is great pressure, due to the increasing
mathematization of society and the job marketplace, to
introduce new topics into the curriculum. At ICME IV
the most talked about areas for such topics were the
newer applications, calculators, statistics, and computer literacy. In general, the total time available
to mathematics within a given school's curriculum is
constant, so inserting these topics forces less time to
be devoted to some topics now in the curriculum. Yet
the time devoted to a particular concept has been shown
to be a crucial variable in predicting student performance on that concept (see [3], [4]). Thus, if we are
to modernize the curriculum, some existing topics or
expected competencies will have to be dropped or given
less emphasis. Common candidates for exclusion or
decreased attention have been paper and pencil arithmetic, unrealistic word problems, Euclidean geometry,
computational logarithms, numerical trigonometry, sets,
trinomial factoring, and proof.

In the United States, decision-making is done at
the local school district or the state level, and
reports of commissions of national scope are looked to
only for general guidance rather than as stating any
sort of regulatory policy. The past five years have
seen four reports ([5], [6], [7], and [8]) dealing with
the K-12 curriculum as a whole, after a twelve-year
hiatus since the last report of this kind [9]. This
flurry of reports reflects the concern of the U.S.
mathematics education community towards today's curriculum. Accordingly, these reports all contain many
recommendations. Regarding content selection, each
report lists what it would like to see in great detail.
Yet the reports give at most brief attention to the

question of what should be deleted. It was the lack of evidence of visible concern for this latter question that led to the organization at ICME IV of a panel entitled "What Should be Dropped from the Secondary School Curriculum to Make Room for New Topics?"

Because my opinions on this topic reflecting specifics of the United States situation are already available (in [10]), I will focus here only on general issues regarding the selection of content for the curriculum.

Why is Content Taught?

Curriculum tends to be self-perpetuating because it is easiest to teach what you have already taught or what you have been taught. For this reason, many people, when asked why a given topic is to be taught, respond in ways that beg the question,

> "It's in the book. The people who wrote the book know more than I do. I should follow them."

> "It's in the syllabus. I have no choice."

> "I had it when I was in school and I've turned out just fine."

or that use false tautologies:

> "Hard work is good. To learn concept X requires hard work. Therefore concept X is good."

> "A general area (such as algebra) is important. Therefore any specifics relating to that area (such as trinomial factoring) are important."

> "It's good for the students who are best. Therefore it's best for all students."

More thoughtful responses tend to fall into four categories:

A. Importance to life outside of school
   1. The concept is important for a consumer to
      know--i.e., it affects everyday life.
   2. The concept is important for an enlightened
      citizen or educated person to know in order to
      make decisions or to select decision-makers.
   3. The concept is useful for getting a job or is
      useful on the job.

B. Importance to individual's later life in school
   4. The concept builds a foundation for other
      important concepts (even though it may not be
      important in its own right).
   5. The concept is found on college or university
      entrance tests, or is basic to a course
      required for college entrance.
   6. The concept appears in a later mathematics
      course; teaching it earlier will help to
      accelerate a student or help the student in
      that later course.
   7. The concept will help in a non-mathematics
      course that applies mathematics.

C. General transfer
   8. The concept helps build important general cog-
      nitive traits such as space perception, criti-
      cal thinking, precision, memory, etc.
   9. The concept promotes desirable moral values or
      discipline, such as neatness, tolerance, uni-
      formity, hard work, avoiding error, etc.

D. Motivation
   10. The concept is fun or encourages students to
       want to learn more.
   11. The concept is easy to learn and builds confi-
       dence.
   12. The teacher enjoys teaching the concept.

   No such short list could cover all the rationales
that are given for teaching mathematical topics. (For
example, "beauty" and cultural aspects of mathematics
are not listed.) Nor is it always easy to identify the
rationale for a given topic; teachers often give reasons for teaching a topic that are different than the
original curriculum determiners had in mind. Yet,

despite the incompleteness of the list, the importance
of mathematics has grown so much in the past generation
that there is much more good mathematics that meets one
or more of the above criteria than can possibly be fit
into the typical student's curriculum. Thus any criterion for inclusion becomes a criterion for exclusion
as well. This forces one to be more critical of the
criteria themselves and to place priorities upon them.

Establishing Priorities

Consider the rationales of Category C. The research has long shown that measurable general transfer
does not occur from studying particular mathematics;
for example, studying proof in geometry does not
improve critical thinking ability (see, e.g., [11]).
Yet teachers often appeal to moral reasons or general
transfer rationale when voicing opinions regarding the
mathematics curriculum [12]. For instance, arithmetic
in the elementary school in today's calculator age is
often defended because it teaches children orderly
thinking, neatness, and precision. The calculator is
viewed as taking away what was never there! In general, transfer rationales seem to be offered when no
other rationales can be found.

By far the most common rationales offered by mathematics teachers come under Category B, importance to a
student's later life in schools. Properties of numbers
are said to be taught because they are useful in early
algebra; the selection of content for early algebra is
based upon its utility in later algebra; later algebra
is important for calculus; and so on. When a substantial segment of the student population will have such a
later life, these rationales are without question
valid. But how many of today's arithmetic students
will go on to take calculus? If the factoring of poly-
nomial expressions in first-year algebra is done so
that two or three years later a student will understand
the Factor and Remainder Theorems, we must ask how
important these theorems are and for what reason are
they in the curriculum. We must ask if these
rationales are sufficient in an age when topics are
competing for space on an overcrowded list of potential
interesting content.

In the United States it is commonly felt that the college entrance tests known as the SAT's (Scholastic Aptitude Tests) are based upon the secondary school curriculum despite reports by the test creators that the tests are designed to be relatively curriculum-free. Appeal to such tests seems only to be a disguised way of begging the question: if one cannot think of a good reason for teaching a topic, then assert that the topic is necessary for performance on some test which was designed after the topic had found its way into the curriculum.

The compelling reasons for inclusion or exclusion of a topic from the curriculum are those of Category A: the topic's importance to the student as consumer, enlightened citizen, or worker. In an earlier age, these rationales would not have been enough to insure mathematics its important position in the curriculum. Frankly, the typical consumer encountered only arithmetic and the simplest geometry of measure, the enlightened citizen needed to know little more than that, and few workers used mathematics above trigonometry on their jobs.

Today, however, the consumer is faced with many personal decisions that can most easily be analyzed using elementary algebra and geometry. The enlightened citizen is bombarded with statistical information, and the worker comes into increasing contact with the language of computers, linear algebra, and analysis. Moreover, with the ubiquity of calculators and the increasing presence of computers in the home, the future seems even brighter for those of us who enjoy the beauty and the power of mathematics. This importance carries with it the responsiblity to select that mathematics which is most useful to the majority of our students and delete that mathematics which is in the curriculum for other reasons.

It remains to consider the rationales under motivation, Category D. If content cannot be motivated, i.e., if a student cannot be convinced of the value of a particular bit of content, then there is reason to question whether that content should be in the curriculum. Yet it is exceedingly important that

students like the mathematics they take. Thus it
becomes the job of mathematics educators--perhaps the
most important job of mathematics educators--to take
that mathematics which is deemed to be valuable (ignoring its motivational value) and present it in a way
which will benefit the student most in the long run
while at the same time encouraging the student to want
to study more in the short run. There is too much
mathematics which is important today to allow us the
luxury of topics that remain in the curriculum merely
because teachers like to teach them. Thus while motivation is necessary in presentation of content, the ability to motivate a presentation should not be an argument for selection (except as all other criteria are
equal).

We are left to conclude that importance for life
outside school is the most valid criterion for content
selection in mathematics. Of course, this discussion
and its conclusion oversimplify the content selection
issue in a number of ways. Sometimes selection is not
"all or none" but a matter of degree. For example, it
seems important to teach mathematical modeling to
secondary school students at some time, but it is not
easy to determine how much is needed and when it should
be taught. Sometimes students should only have a
glimpse of a topic, to help them select interests. Yet
when we teach only a glimpse, as has often been done
with sets or groups or vectors or matrices, there is so
little payoff that many who study them think of them
afterwards as less important than before they had studied them.

Sometimes the real world or job market is changing
so quickly that content selection cannot be easily
done. For example, what programming language should be
taught? And to what extent should we employ iterative
processes, as used by computers, to solve quadratic
equations? Sometimes our selections of content are
correct but our approaches are faulty. We need to
teach that a topic isn't automatically made more useful
by being taught in a more rigorous or more abstract way
nor by being presented in a less formal way. The best
approach varies from student to student, teacher to
teacher, and topic to topic.

Finally, the best content for a given student
varies among students even of the same ability and
background, and most students are too little familiar
with mathematics to make wise choices about this content, forcing mathematics educators into the hard reality of having both approaches and content which are
certain not to be optimal for all students all of the
time. What should be or not be in the mathematics curriculum is thus a crucial problem whose solution can
always be improved.

Who Is Listening?

For discussion of curriculum deletion issues, a
conference such as ICME IV provides an optimal setting
because, for almost any bit of content one wishes to
delete, there exists some place in the world where that
content has never been taught or has already been taken
out of the curriculum. Hearing from these places, one
can imagine possible results of a decision before it is
made, a valuable asset to have if you are a decision-
maker. In this sense, the Congress was potentially
most valuable.

However, fruitful discussion of these issues is
still needed and is difficult to undertake unless those
who disagree are brought together for discussion. Yet,
at conferences like these, sessions that discuss
geometry are attended almost exclusively by those who
are positively inclined towards the place of geometry
in the curriculum; sessions that discuss statistics are
likewise attended by those conversant with and favor-
able towards statistics; and so on for computers, or
calculus, or any other subject. Thus, to generate
spokespersons for both sides of an issue, the issue
itself must be known in advance and speakers must be
solicited to respond to a particular side of the issue.

It might not be easy to find a distinguished person to address a side of an issue unpopular to mathematics educators. Yet that side might well be the more
popular side to those outside of mathematics. For
example, the importance of computers seems to enhance,
rather than detract from, the importance of other
mathematics. So it is not at all obvious that time

should be taken from other mathematics to devote to
instruction with or about computers. Perhaps mathematics educators should be arguing in schools and among
governmental agencies for more total time to be devoted
to mathematics. It would be difficult to find someone
in mathematics to publicly disagree with this position.
But it would be easy to find someone outside of mathematics who would argue against giving mathematics more
time. We should be ready to listen to speakers with
whom we disagree.

Having dissenters at a conference increases communication in two directions: those on each side are
made aware of the complexity of a given issue. For
example, many speakers at ICME IV spoke as if the
widespread use of calculators in the elementary school
classroom was a foregone inevitability. (Two decades
ago people were predicting the same for television.)
Yet research in the United States shows that the typical elementary school teacher opposes the use of calculators in the classroom [14]. Ignorance is surely not
the only reason for this opposition. Yet we will not
be forced to confront the other reasons until we bring
in those who disagree; they in turn will not realize
the strength of our arguments until they have to
respond to them.

Where Were the Decision Makers?

We might characterize this situation by saying
that "the society of dissenters" was absent from the
Congress. For the United States, even more noticeably
absent was a "second society," the society of those
people who make the decisions about the actual mathematics curriculum to be received by students: textbook
publishers, state supervisors, and individual teachers.

For example, my wife is an executive editor in
mathematics (K-12) for a textbook publisher. No significant curriculum change can today be made without the
cooperation of publishers. Yet we believe she was one
of only three employees of publishers' editorial
departments in mathematics at ICME IV from the entire
United States. State mathematics supervisors from only
three of the fifty United States were represented. Not

one mathematics teacher or department chairman from any
pre-college institution in our home area of Chicago was
in attendance, even though this area of approximately 8
million people tends to be quite active in mathematics
education nationally.

The corresponding decision-makers from other countries with more centralized educational organizations
were present at ICME IV. Potentially this has sad
consequences for the immediate future of mathematics
education in the United States. We, who as much as any
society led the revolution in school mathematics of two
decades ago, whose computer technology has been the
impetus for many of the changes now being recommended
throughout the world, and whose research community is
by far the largest of any country, seem poised to
quickly fall behind the rest of the world in the quality and relevance of the mathematics which we teach to
the majority of our students. Just as the United
States steel industry became complacent while others
modernized their industries after World War II, and
just as the United States auto industry smugly promoted
large gas-guzzling cars and refused to go to small cars
despite warnings of gas shortages, the typical United
States mathematics teacher is quietly teaching loads of
content that will be appropriate only for future mathematics teachers, ignoring both the advances in mathematics and in its applications that have occurred in
the past thirty-five years. There was a lot said at
ICME IV, but it was said to those who are not responsible for curriculum decision-making in the United
States.

Notes and References

[1] This suggestion does not appear in the Report of
    the Conference on the K-12 Mathematics Curriculum,
    Snowmass, Colorado, June 21-24, 1973 (Bloomington,
    IN: Mathematics Education Development Center,
    Indiana University, 1973), because recommendations
    in that report required some group consensus.
[2] For information, write the Calculator Information
    Center, 1200 Chambers Road, Columbus, Ohio 43212.
[3] Barak Rosenshine and David Berliner. "Academic

engaged time." British Journal of Teacher Education 4 (1978) 3-16.

[4] David Berliner. "Allocated time, engaged time, and academic learning in elementary school mathematics instruction." Paper presented at the National Council of Teachers of Mathematics annual meeting, San Diego, California, 1978.

[5] National Advisory Committee on Mathematical Education. Overview and Analysis of School Mathematics: Grades K-12. Washington, D.C.: Conference Board of the Mathematical Sciences, 1975.

[6] National Council of Supervisors of Mathematics. "Position paper on basic mathematical skills." Minneapolis, Minnesota: NCSM, 1977.

[7] "NCTM-MAA position statement on recommendations for the preparation of high school students for college mathematics courses." Washington, D.C.: NCTM or MAA, 1978.

[8] National Council of Teachers of Mathematics. Agenda for Action. Washington, D.C.: NCTM, 1980.

[9] Cambridge Conference on School Mathematics. Goals for School Mathematics. Boston, Massachusetts: Houghton Mifflin, 1963.

[10] Zalman Usiskin. "What should not be in the algebra and geometry curricula of average college-bound students?" The Mathematics Teacher 73 (September 1980) 413-424.

[11] Harold Fawcett. Thirteenth Yearbook: The Nature of Proof. New York: New York, National Council of Teachers of Mathematics, 1938.

[12] Robert E. Stake, Jack A. Easley, et al. Case Studies in Science Education: Volume II: Design, Overview and General Findings. Washington, D.C.: National Science Foundation, 1978, pp. 12-33 ff.

[13] James Braswell. "The college board scholastic aptitude test: An overview of the mathematical portion." The Mathematics Teacher 71 (March 1978) 168-180.

[14] National Council of Teachers of Mathematics. Priorities in School Mathematics. Final Report, April 1980. Columbus, Ohio: ERIC Center for Science, Mathematics, and Environmental Education, Document Nos. SE 030 577, SE 030-578.

Major Trends from ICME IV:
A Southeast Asian Perspective

Bienvenedo Nebres
Ateneo de Manila University

In our work in mathematics education in Southeast
Asia, we have become aware that there is a dual aspect
to our relationship with developments in mathematical
education in the advanced countries. On the one hand,
we are aware of the great differences in our situation
(thus there is need of great care in adapting to new
trends); on the other hand, we cannot help but be
affected by new developments in the West (because our
textbooks are adapted from the West, and our young
mathematicians are trained there).

Differences

The first impression of a visiting mathematics
educator from the United States or Europe in discussions with Southeast Asian counterparts is that of
similarities with the Western situation. Our students
also suffer from versions of "math anxiety," teachers
are poorly trained and even more poorly paid, materials
in the syllabus are not adequately covered. As he begins to discuss possible solutions, however, he begins
to realize that beneath these similarities are greater
differences. Many of the solutions developed in
advanced countries to help develop functional numeracy
or relieve "math anxiety" cannot apply to a country
such as the Philippines, where the textbook ratio (all
textbooks, not just math textbooks) is one textbook for
every ten students. In-service programs for teachers

are extremely difficult to implement in a situation
where teachers have to teach thirty or more hours a
week in order to earn a living. There is the additional burden of language. Most Filipino children
speak one language at home, then have to learn both
Pilipino and English when they go to school. Half
their subjects are taught in Pilipino, half in English.

In efforts to improve mathematical education in
developing countries, we must therefore be aware of the
human and material infrastructure underlying this task.
E.F. Schumacher has an eloquent passage ([1], p. 164-
165) about the transplanting not of methods of mathematical education, but of a great refinery. The evolution of this refinery began with something simple;
"then this was added and that was modified, and so the
whole thing became more and more complex." But even
more, this complex establishment is but the tip of an
iceberg. We cannot see "the immensity and complexity
of the arrangements that allow crude oil to flow into
the refinery and ensure that a multitude of consign-
ments of refined products reaches innumerable consumers
through a most elaborate distribution system. Nor can
we see the intellectual achievements behind the planning, organizing, the financing and marketing. Least
of all can we see the great educational background
which is the precondition of all, extending from primary schools to universities and specialized research
establishments...the visitor sees only the tip of the
iceberg."

The dominant reality about the underside of the
iceberg in developing countries in the scarcity of
resources, both human and material. On the human side,
there are few experts who can be expected to carry out
the work of improving mathematics education and these
few are usually overextended in other tasks. Teachers
have to work long hours for little pay. In many countries, the intellectual and educational tradition is
young (and weak), so that the support expected from
home, from government and from society is often lack-
ing. On the material side, textbooks, classrooms,
blackboard and chalk are scarce. Students have to cope
with several languages, and most drop out before com-
pleting four or six years of school because of the

pressures of poverty. It is within this social matrix
that trends and developments in mathematics education
must be adapted. It is important, therefore, to understand the (human and material) conditions behind these
trends and developments. We have to work on these
preconditions as well, otherwise our transplants will
wither in this unaccommodating soil.

Influence of New Trends

The paradoxical aspect of these differences
(especially the scarcity of human and material resources) is that instead of isolating us from developments in advanced countries, they make us more vulner-
able to them. This is because we have to depend on
Western mathematics educators and Western textbooks.
We do not have the necessary number of experts nor the
funds to develop our own textbooks. So our textbooks
are either completely Western or are minor adaptations
of Western textbooks. In some countries, teams of
Western experts have been invited to write their textbooks. Similarly, we depend on outside experts for our
in-service programs for teachers, for advice on trends
and recent developments.

It is the educational version of the well-known
problem of economic colonialism. We import jeans and
Pepsi-Cola, then we learn to produce them ourselves and
are pleased at how advanced we have become. We mistake
the trappings for the substance. Thus, some teachers
on the advent of the new math to the Philippines happily wasted hours teaching their pupils how to make
nice curly brackets, how to count in Chinese, in base
eight, and so forth, and never quite got to teaching
them how to add and subtract.

This essay is a reflection mostly from the point
of view of the Philippines and Southeast Asia on main
trends from ICME IV. I choose, of course, the trends
that impressed me most and which I think will have the
greatest impact on our situation. These trends centered on the role of basic skills, the centrality of
problem-solving in mathematics education, and the role
of calculators and computers.

Reflections on the New Math

As the back-to-basics movement gathers momentum in developing countries, it is useful for us to reflect on the successes and failures in the earlier implementation of the new math. Some lessons in this area apply as well to present efforts at return-to-basics. It is useful also to know more precisely what the imbalances are that we are retreating from.

There were two streams of implementation of the new math in the Philippines in the early 1960's. The first one might call the elite stream. This was implemented in a few private schools, based on the Addison-Wesley series, and handled by experts from college mathematics departments. It involved continuing training of the grade school and high school teachers from these select schools over a period of several years. This was quite successful and has produced very good students over the past years.

The second would be the popular stream. In the early 1960's, the Education Ministry in the Philippines launched a massive campaign to introduce the new math to all public and private schools. Assisting them were Peace Corps volunteers from the United States. But the program extended itself too far, too fast. The so-called experts were not so expert and confused fads with substance; textbooks were in short supply, the training periods inadequate. The end-result were teachers who identified the new math with curly brackets and exotic number systems.

The spiral method used in the design of the new math curriculum also produced unforeseen consequences. The method followed the same sequence of topics every year--sets, set operations, number systems, operations on numbers--in growing complexity. It is a well-known phenomenon that teachers tend to cover the first few topics in the curriculum thoroughly and never have enough time for the final topics. Thus in later years, we found students who thought that mathematics was sets, set complements, one-to-one correspondence, who

could not add and subtract two-digit numbers. As we
implement the return-to-basics movement, we might
reflect on our past experience and on the necessary
pre-conditions for successful educational reform on a
massive (national) scale.

Balancing the Why and the How

The new mathematics emphasized concepts and
formalization (definitions and axioms, knowing why)
versus skills and intuition (mastery of basic operations, problem-solving, knowing how). If we were to
illustrate this with a lesson in addition, to add 199 +
87, things might go as follows:

Old Math		New Math
199	9	$199 = 1\cdot100+9\cdot10+9$
87	7	$87 = 8\cdot10+7$
———	—	$199+87 = (1\cdot100+9\cdot10+9)+(8\cdot10+7)$
	6 carry 1	$= 1\cdot100+(9\cdot10+8\cdot10)+(9+7)$
		$= 1\cdot100+(9+8)\cdot10+16$
1	11	$= 1\cdot100+17\cdot10+1\cdot10+6$
99	199	$= 1\cdot100+1\cdot100+7\cdot10+1\cdot10+6$
87	87	$= (1+1)100+(7+1)\cdot10+6$
———	———	$= 2\cdot100+8\cdot10+6$
86 carry 1	286	$= 286$

The old math was criticized for making things look
magical. We learned "carrying" by rote memorization
without understanding. On the other hand, overemphasis
on the why of things often ended up with no time left
for children to learn how to add and subtract. It is
clear now in retrospect that it is useful to give the
new math explanation of "carrying" (explaining posi-
tional notation, application of laws governing basic
operations). This could be given with a few simple
examples, before one teaches "carrying." But it is not
a method, not an algorithm. The algorithm is given by
"carrying."

Return-to-basics asks then not for a return to
rote memorization without understanding, but for a
better balance between learning the meaning of what we

are doing and actually acquiring the (mechanical)
skills to do them well. We must balance the why and
the how.

A Sense of Algorithm

If we analyze the addition algorithm (say, the
addition of two whole numbers), we see that the algorithm consists of a memory part (learning the addition
table for 0 to 9) and a mechanical procedure part (carrying). Carrying is simply a procedure for reducing an
addition problem for numbers greater than 10 to an
addition problem for numbers less than 10. Most
mechanical skills that we teach in school mathematics
are algorithms of this sort: they have a memory part
(operations on relatively simply objects) and a mechanical procedure part (to reduce operations on more complex objects to operations on simpler ones).

The advent of calculators and computers have made
us much more aware of the need for a sense of algorithms. It may be useful to ask whether our problem in
teaching fractions may not be due to the fact that we
have not developed suitable algorithms for operations
on fractions. Or perhaps we do not have a suitable
algorithm for the teaching of these algorithms. The
use of calculators also demands the development of new
algorithms. How do we use calculators to teach operations on fractions?

Keeping Fads in Perspective

Another lesson I hope we have learned from the
new math period is to distinguish between solid innovations and fads. The fashion then was set-theory and
foundations, thus the fascination with sets, numbers
and numerals, ordinal and cardinal numbers, exotic
number systems, one-to-one correspondence and so forth.
It is difficult to believe now that we actually tried
to teach the distinction between a number and a numeral
to grammar school students.

In the Philippines, at least, part of the blame
for these mistakes falls on the mathematicians in colleges and universities for either fostering these

oddities or not being interested enough to help teachers in grade schools and high schools keep things in perspective. We believe now that many of the excesses could have been kept under control had there been better cooperation between university and school mathematics. Present efforts at reform have their own emphases and buzz-words: basic skills, calculators, estimation, problem-solving for life. Bringing these concepts and thrusts into the day-to-day work of the classroom without distorting them badly will require much cooperative work from the whole mathematics community.

Basic Skills

What should we mean by basic skills? Do we mean merely computational skills? We mean these and more. The Agenda for Action of the National Council of Teachers of Mathematics gives an excellent presentation ([2], p. 7) of what we should mean by basic skills (beyond computational facility):

> There should be increased emphasis on such activities as:
> -- organizing and presenting data
> -- interpreting data
> -- mentally estimating results of calculations
> -- using technological aids to calculate
> -- using imagery, maps, sketches, and diagrams as aids to visualizing and conceptualizing a problem.
>
> There should be decreased emphasis on such activities as:
> -- isolated drill with numbers apart from problem contexts
> -- performing paper-and-pencil calculations with numbers of more than two digits
> -- mastering highly specialized vocabulary not useful later in mathematics or in daily living.

These recommendations are certainly very reasonable and at first sight it may appear that there should

be no serious problems with their implementation. But
especially when taken with the further recommendations
regarding the centrality of problem-solving and the use
of calculators, it becomes clear that they demand much
rethinking regarding the day-to-day content and structure of the mathematics class. We shall return to
these implications for the classroom after discussing
these further recommendations.

Problem-Solving

The emphasis on the centrality of problem-
solving has probably different meanings for different
people. My own first reaction was a sigh of relief
that we were getting over the period of emphasis on
soft, definitional type mathematics and getting to
hard, problem-solving mathematics. But the emphasis of
ICME IV and the NCTM recommendations mean much more
than that. They talk about problem-solving for life
([2], p. 2):

> Mathematics programs should give students
> experience in the applications of mathemat-
> ics, in selecting and matching strategies to
> the situation at hand. Students must learn
> to:
> -- formulate key questions;
> -- analyze and conceptualize problems;
> -- define the problem and the goal;
> -- discover patterns and similarities;
> -- seek out appropriate data;
> -- experiment. [2, p. 2]

These are deep and far-ranging goals, worthy of our
best efforts in the 1980's. But they will require
rethinking, reflection, and change in our accustomed
manner of teaching.

Necessary Changes

Our present teaching pattern organizes material
into bite-size pieces (each piece suitable for one or
two class periods). However, learning the kind of
problem-solving approach envisioned above requires the
ability to understand a whole complex situation, to

break it down into manageable parts, and then to solve
it. To have the situation pre-broken-down or pre-
analyzed defeats the whole learning process. We may
have to learn to adopt something of a case-approach or
a project-approach. But who will develop the cases and
the projects? What will this mean for the development
of textbooks?

Likewise, the kind of problem-solving material
envisioned above cannot be easily adjusted to class
periods consisting of 40-50 minutes three or four times
a week. We will need longer stretches of time, as in a
science laboratory or in a case study, for the analysis
of complex problem situations. Mathematics may have to
become something like a lab science, with regular class
periods together with longer stretches for problem
analysis and discussions.

Above all, we as mathematics teachers will have to
learn to see mathematics not as a collection of methods
and tricks to solve (once and for all) certain pre-
ordained textbook problems, but as a science of understanding, analyzing and at least partially solving some
of the more important problems around us. Solutions
will be less of the neat, closed type, and more of the
approximate, probabilistic kind. I believe the implications of this need for change will have a far-
reaching influence on the content and process of our
teacher-training courses.

Priorities in Material and Time

These proposals for reorganization of material
and time inevitably present us with the problem of
allocation of time. There is so much to be learned and
so little time. There is a need, first, to specify
more sharply what we mean by basic skills, not just
listing them, but giving them in some order of priority. Can we specify these skills in categories of
"absolutely necessary", "useful", "desirable if you
have time"? We learned this need from the implementation of the new math. Without such guidelines, teachers may cover the useful and desirable, yet miss the
absolutely necessary. Is it possible, further, to give
a breakdown of such basic skills grade level by grade

level, with diagnostic tests at the beginning of each
new grade level to see if the earlier skills have been
sufficiently mastered? Without such hard norms, our
experience in the Philippines is that goals remain
aspirational and rhetorical, not real.

Second, time constraints require that we review
the teaching of computational skills. On the one hand,
handheld calculators are more readily available. On
the other hand, the mastery of many routine skills
requires much time.

> Insisting that students become highly facile
> in paper-and-pencil computations such as 3841
> times 937 or 72,509 divided by 29.3 is time-
> consuming and costly. For most students,
> much of a full year of instruction in
> mathematics is spent on the division of whole
> numbers--a massive investment with increas-
> ingly limited productive return. A small
> fraction of the time is spent on the skills
> of problem analysis and interpretation, which
> enable students to identify and set up the
> computations needed. For most complex prob-
> lems, using the calculator for rapid and
> accurate computation makes a far greater con-
> tribution to functional competence in daily
> life. [2, p. 6]

This recommendation envisions a situation where we
    -- teach understanding of basic operations using two-
       or three-digit numbers;
    -- show how computations with larger numbers are done
       on a calculator;
    -- drill the students in estimating expected answers
       so they can see if the calculator answer is within
       the expected range.

Educators in developing countries may argue that
most of our students cannot afford calculators (many of
them do not even have a single textbook to their name).
But we should not give up on this recommendation so
readily. Many of our poorer students will need these
calculation skills in future job situations. It may be
possible to set up "calculator rooms" in schools, where

students can spend time learning the use of calculators
or doing their homework exercises. It is also important to remember the reason for this recommendation: to
give more time to the teaching of problem analysis and
interpretation. The need for this training is perhaps
even more urgent in the developing countries.

Third, there is need to devote more time in absolute terms to the study of mathematics in grade school
and high school. There has been an unfortunate ten-
dency in many developing countries, including the Philippines, to introduce more and more new courses
(social studies, environment, etc.) in the early years.
Much time is also taken from the classroom by so-called
civic duties. At an education conference held four
years ago, a committee of which I was chairman decided
that the most important reform we could recommend was
simply to give back to teachers the time to teach and
to students the time to learn. The NCTM recommends a
minimum of five hours a week in the first four years
and seven hours a week in the next four. The study of
mathematics requires substantial effort and time. A
good teacher can make the content clearer and our time
more interesting. He cannot substitute for effort and
time.

A Final Note

The implementation of the major reform trends
from ICME IV (as well as the NCTM recommendations) will
clearly require much effort, time, and material
resources. Are they worth such effort and such
resources? This question has a particular urgency in a
poor developing country such as the Philippines, where
our need for resources has to complete with such basic
needs as food, housing, and primary health care. In
recent years our government has given primary attention
to the development of material resources (infrastructure, energy) over human resources (education, health).
But these priorities are short-sighted. The example of
other countries is clear. There are many countries
with extremely limited natural resources, but with a
highly skilled populace. They prosper. There are
other countries with immense natural wealth, but whose
people are in large part poorly educated. They have a

few wasteful rich and a majority in poverty and desti-
tution. Even for the development of our own natural
resources our biggest handicap is not financial but
human: the quality of our management and entrepeneur-
ship.

The lesson is clear: if we continue our present
lack of emphasis on the development of our human
resource, our natural wealth will be enjoyed by a rich
few and the foreign experts who have the skills to
develop our resources. Education remains our main tool
for the development of people. Together with language,
mathematics holds a privileged place in the educational
system. To our earlier question then we answer that
these reforms are worth such efforts and such
resources. Better mathematics education is urgently
needed for our younger generation to understand and
cope with the modern world.

References

[1] E.F. Schumacher. Small is Beautiful. Harper and
     Row, 1975.
[2] An Agenda for Action. National Council of Teach-
     ers of Mathematics, 1980.

Features

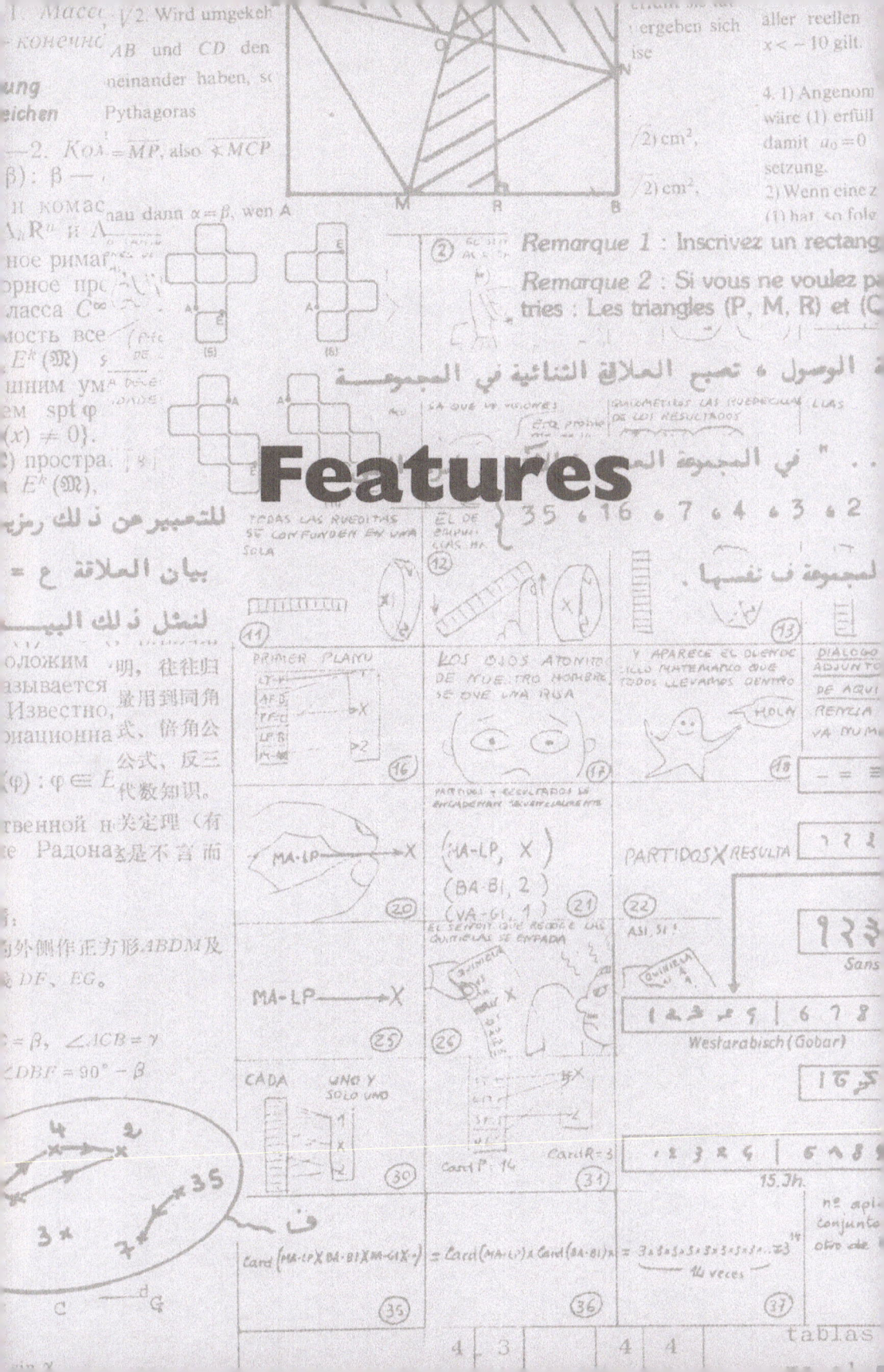

George Pólya

The organizers' choice of George Pólya to be
Honorary President of the Congress was unusually timely
and appropriate. How was it possible to predict the
tone of the speakers at the plenary sessions or the
concern shown in various sessions over the use of problems in teaching and over the choice of real problems
from science? Pólya has been telling us for years
about the effectiveness of carefully chosen problems in
mathematics teaching and the use of intuition, guessing, and real world experience in teaching important
mathematical ideas.

In a recent interview [1] Pólya pointed out that
his interest in problem solving was related to his own
experience in choosing and solving problems in his
research. But his commitment to teaching through the
use of problems was first evident in the now classic
set, written with G. Szegö, the <u>Aufgaben</u> <u>und</u> <u>Lehrsatze</u>
<u>aus</u> <u>der</u> <u>Analysis</u> (<u>Problems</u> <u>and</u> <u>Theorems</u> <u>of</u> <u>Analysis</u>) of

1925. This work set out to convey to the reader,
through carefully chosen and graded problems, without
accompanying text, knowledge of a number of areas of
mathematical analysis. This was an essentially new
idea.

It was in How to Solve It, however, that he wrote
down a set of rules and guidelines for tackling mathematical problems. Again, in the same interview, he
pointed out that an earlier, unpublished version in
German, written when Pólya was still at the Federal
Institute of Technology in Zürich, was less successful
than the version finally published in English in 1945.
Pólya came to the United States in 1940. When he told
G.H. Hardy that he was writing a book How to Solve It,
Hardy remarked, "It is appropriate that you go to America. It is the country of 'How to...books'."

HOW TO SOLVE IT

G. POLYA

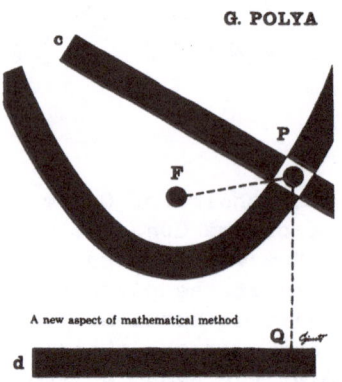

A new aspect of mathematical method

Pólya went to four publishers before finding one
willing to publish How to Solve It. How would those
editors who turned down the manuscript feel today,
knowing that well over a half million copies have been
sold in English and that it has been translated into 16
languages? [2]

The ideas in How to Solve It were expanded and
made more specific in his subsequent Mathematics and
Plausible Reasoning, Mathematical Discovery and
Mathematical Methods in Science, as well as in numerous
journal articles and several films.

Though Pólya had illustrious antecedents--Descartes, Leibniz, Euler, and Hadamard, to name a few--who tried to analyse the process of problem solving, it is, nevertheless, largely Pólya whom we think of today when we systematically look at approaches to problems: looking at concrete examples, checking for patterns, trying extreme cases, looking for analogous problems, specializing and generalizing, and so on. And it is Pólya who has argued for years for concrete problems, chosen often from the sciences, that will have meaning for the student, that are chosen for the correct age level, and for which a student has sufficient background to understand. If a student is to solve a problem, he should care about the answer.

It is heartening today to find so much interest in Pólya's teaching evident in this Congress and, indeed, elsewhere. One need only cite the publication of the 1980 NCTM yearbook Problem Solving in School Mathematics, the NCTM's first recommendation in its new curriculum study [3] that "problem solving be the focus of school mathematics in the 1980's," the appearance of compendia of applied problems from CUPM (the Committee on the Undergraduate Program in Mathematics of the Mathematical Association of America) and elsewhere, the call for more use of problems in teaching by the new editor of the American Mathematical Monthly, Paul R. Halmos [4], and the work of UMAP (Undergraduate Mathematics and Its Applications Project), as a few examples that Pólya's counsel is being heeded.

References

[1]   Gerald L. Alexanderson. "George Pólya interviewed on his ninetieth birthday." Two-Year College Mathematics Journal 10 (1979) 13-19.

[2]   How to Solve It has been translated into Arabic, Croatian, Dutch, French, German, Hebrew, Hungarian, Italian, Japanese, Polish, Portuguese, Rumanian, Russian, Slovenian, Spanish and Swedish.

[3]   Agenda for Action, National Council of Teachers of Mathematics, 1980.

[4]   Paul R. Halmos, "The heart of mathematics," Amer. Math. Monthly 87 (1980) 519-524.

Pólya in Action

The problem has been posed by Pólya and his student assistants to a group of teachers at the
annual Asilomar Meeting of the California
Mathematics Council.

The solution has been happily discovered.

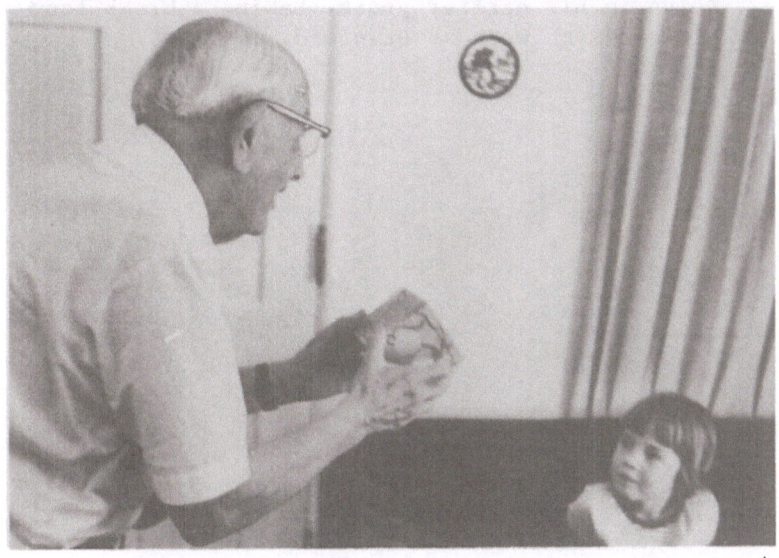

Pólya greatly enjoys children. At faculty picn-
ics, he usually can be found surrounded by youngsters. Here, at age 90, we see Pólya giving
five-year old Lisa a playful lesson on polyhedra.

Pólya in his earlier years, taking a break from
his research work at Stanford.

Pólya and his wife Stella.

Minicourses

The emphasis at ICME IV on applications, problem solving and mathematical modelling led the program committee to include special minicourses to introduce participants to some of the newer topics that may become part of the school curriculum of the 1980's. Here are vignettes from four of these minicourses, based on material provided by J.H. van Lint (Coding Theory), Ronald Graham (Ramsey Theory), Richard Fateman (Computer Algebra), and Ram Gnanadesikan (Exploratory Data Analysis).

Coding Theory

In modern communication theory messages are often sent over a channel (telephone, telex, satellite link) as a string of two different kinds of symbols (which we call 0 and 1). Due to technical deficiencies these channels are often "noisy," the effect being that the receiver interprets a 0 signal as a 1 (or vice versa) with a certain (small) probability p. Algebraic coding theory studies the problem of using these channels economically in such a way that the probability of error at the receiver is minimized.

Suppose, as an example, that we have an alphabet of only four letters, say a,b,c,d, which we <u>code</u> as a = 00, b = 01, c = 10, d = 11. To prevent errors, we repeat each symbol three times when transmitting it, so b is transmitted as 000111, for example. Clearly the receiver can recognize and correct any single error by using a 2 out of 3 rule for interpreting. (For example, the received message 010111 would be decoded as b.) However, this is not the most efficient scheme because the code a = 00000, b = 11100, c = 10011, d = 01111 achieves the same goal yet the <u>words</u> of this code consist of only five 0's and 1's instead of six. To see this, observe that any two words differ in at least 3 places (we say their <u>distance</u> is greater than or equal to 3). Therefore a word with one error looks more like the original word than like any other word.

Here is a typical recent problem. To transmit a picture of Saturn to Earth, it is divided into a grid of little squares and for each of these the blackness is measured on a scale from 0 to 63. All numbers are expressed in binary notation (0 = 000000, 1 = 000001, ..., 35 = 100011, ..., 63 = 111111), which helps to explain why the intensity scale ranges from 0 to 63. As in the example above, but in a more sophisticated manner, coding theory is used to devise a systematic way of lengthening each string of six 0's and 1's to 32 symbols in such a way that a received codeword (of "length" 32) which contains not more than seven errors will still be correctly interpreted. A second aspect of this problem, equally important, is to design a fast algorithm for the receiver to correct errors, that is, to <u>decode</u> the message.

A second typical problem in coding theory is to investigate the maximum number of words of length n with mutual distance at least d, which is called A(n,d). For example, A(6,3) signifies the largest number of words of length 6 in which any two differ in at least 3 places. It is easy to verify A(6,3) < 9. By finding an example of 8 words satisfying these conditions, one can prove that A(6,3) = 8. Many areas of combinatorial theory play a crucial role in this part of coding theory.

References

[1]   F.J. MacWilliams and N.J.A. Sloane. The Theory of Error-Correcting Codes. North Holland, 1977.

[2]   J.H. van Lint. Introduction to Coding Theory. Springer-Verlag, 1981.

[3]   P.J. Cameron and J.H. van Lint. Graphs, Codes and Designs. Cambridge University Press, 1980.

[4]   N.J.A. Sloane. A Short Course on Error Correcting Codes. CISM Courses and Lectures 188, Springer-Verlag, 1974.

Ramsey Theory

Everyone knows that if the ordinary game of tic-tac-toe is played carefully, the game will always result in a draw. It is not as well known that 3-dimensional tic-tac-toe played on a 3 x 3 x 3 board, can never result in a draw. Problem: Try to prove this.

Tic-tac-toe is a simple example of a structure governed by Ramsey Theory, a very exciting and active area in combinatorics. The essence of Ramsey Theory is captured by the slogan "complete disorder is impossible." More precisely, if a structure is large enough, it must contain a substructure having a predictable amount of order. Here are some more examples:

1.   Any rearrangement of the integers 1,2,...,101 must contain either an increasing subsequence of 11 terms or a decreasing subsequence of 11 terms. (Why?) More generally, how many numbers are needed if we want to be guaranteed an increasing or

decreasing sequence of length 100, or of length n?
(Answer: $n^2 + 1$).

2. For any 6 people, there are always 3 who either all
    know each other or are all mutual strangers. Prob-
    lem: How many people must be taken to guarantee
    that we can always find 4 such people? (Answer:
    18.) No one knows the answer for 5. In fact, this
    seems likely to remain unknown for the foreseeable
    future.

3. Any set of 5 points in the plane (no three on a
    straight line) contains 4 points which form a con-
    vex quadrilateral. (Easy.) Similarly, any set of 9
    points contains 5 points which form a convex penta-
    gon. (Harder.) Question: How many are needed if
    we want a convex n-gon? Answer: No one knows! 2^{n-2}
    + 1 is enough, and it is conjectured that this is
    best possible. Interestingly, it is not known if a
    convex hexagon containing no other point of the set
    in its interior is ever forced to occur, no matter
    how many points we start with.

4. If the points in the plane are each given one of
    two colors, then there must always be three points
    having the same color which form a $(1,1, \sqrt{2})$ right
    triangle. (Not easy.) This is not true for the
    (1,1,1) equilateral triangle. (Why not?) It is
    not known what happens for the $(1,1, \sqrt{3})$ triangle.
    It is conjectured that all non-equilateral trian-
    gles must occur "monochromatically."

These examples illustrate the type of problems
this branch of combinatorics deals with. They are
often very easily stated, but the solutions can range
from easy observations to great depth. (An example of
a deep result is by Szemeredi's recent theorem that any
infinite set A of integers that contains no k integers
in an arithmetic progression must have "density zero,"
that is, the number of integers in A which are less
than n, divided by n, tends to infinity.)

Suitable topics from Ramsey Theory could form part
of a very attractive introduction to combinatorics for

secondary school students.

References

[1] Daniel Cohen. Basic Techniques of Combinatorial
     Theory. Wiley, 1978.
[2] Martin Gardner. "Mathematical games in which join-
     ing sets of points by lines leads into diverse
     (and diverting) path." Scientific American 237
     (November 1977) 18-28.
[3] Alan Tucker. Applied Combinatorics. Wiley, 1980.
[4] Ronald Graham, Bruce Rothschild, and Joel Spencer.
     Ramsey Theory. Wiley, 1980.

Computer Algebra

A typical computer program for solving the quadratic equation $ax^2 + bx + c = 0$ will require numerical
values of the coefficients, and produce numerical
values of the unknown. For example, if one enters

$$a = 2 \qquad b = -7 \qquad c = 3$$

in such a program, the answer will be

$$x = 3 \qquad \text{or } x = .5$$

if you are lucky. More commonly, it will be something
like

$$x = 2.99998 \qquad \text{or} \qquad x = .49997.$$

But in a symbolic algebraic computer system, one
can enter expressions such as

$$a = q \qquad b = -pq - 1 \qquad c = p.$$

The result is impressive:

$$x = \frac{-(pq-1) \pm \sqrt{p^2q^2+2pq+1-4pq}}{2q} \; .$$

After further exhortation, the program will simplify and reduce this expression, yielding

$$x = p \qquad \text{or} \qquad x = 1/q.$$

Generally, symbolic algebraic programs can manipulate formulas better than a human mathematician. They can solve equations, factor polynomials, differentiate, integrate, compute Taylor series, evaluate determinants, invert matrices and solve differential equations. For example, the program Macsyma will integrate

$$\frac{1}{x^3+2}$$

without complaint, producing

$$-\frac{\log(x^2-2^{\frac{1}{3}}x+2^{\frac{2}{3}})}{62^{\frac{2}{3}}} + \frac{\text{atan}\left[\frac{2x-2^{\frac{1}{3}}}{2^{\frac{1}{3}}\sqrt{3}}\right]}{2^{\frac{2}{3}}\sqrt{3}} + \frac{\log(x+2^{\frac{1}{3}})}{32^{\frac{2}{3}}}$$

in about 50 microseconds of cpu time. If we ask Macsyma to check by differentiating, it disgorges this mess:

$$\frac{1}{3(\frac{2x-2^{\frac{1}{3}}}{32^{\frac{2}{3}}})^2+1)} - \frac{2x-2^{\frac{1}{3}}}{62^{\frac{2}{3}}(x^2-2^{\frac{1}{3}}x+2^{\frac{2}{3}})} + \frac{1}{32^{\frac{2}{3}}(x+2^{\frac{1}{3}})} \; .$$

A bit more urging, called "rational simplification," yields the desired result:

$$\frac{1}{x^3+2}.$$

Until recently these symbol manipulation programs have been available only on large university computers. But recently a concise version called muMath has been developed for the small microcomputers such as Apple II and Radio Schack's TRS-80. This version can easily be used in the average mathematics classroom, providing an unprecedented opportunity for alternative, creative teaching strategies. Microcomputers equipped with programs such as muMath will do for high school and college mathematics what the hand calculator is doing for primary and middle school arithmetic: providing opportunity for appealing realistic problems to replace routine drill and practice, without forcing students to get bogged down in ridiculous detail.

References

[1]   Edward W. Ng (Ed.) Symbolic and Algebraic Computation: Lecture Notes in Computer Science, No. 72. Springer-Verlag, 1979.
[2]   David R. Stoutemeyer. "LISP based symbolic math systems." Byte 4 (August 1979) 176-192.
[3]   David R. Stoutemeyer. "Symbolic math using BASIC." Byte 5 (October 1980) 232-246.

Exploratory Data Analysis

What's the best way to summarize the five numbers 7,8,6,4,100? The average is 25, which is not typical of most (or even any part) of the data. For some

real-life problems 25 would be the proper summary; but
it is often better to summarize the reasonable portion
of the data (7,8,6,4) and to study exceptional values
(like 100) separately to decide, for example, whether
they are interesting special cases, or simply in error.

The median, that number which has half of the data
smaller and half larger than itself, is a _robust_ summary of the data. For this data set it is the middle
value, 7, which is a typical value. Exploratory data
analysis seeks such robust measures as protection
against unusual data and violated assumptions. A few
atypical or "bad" observations can ruin an ordinary
analysis, but will have only a very limited effect on a
robust analysis.

Traditional statistics would use the mean and the
standard deviation to summarize a data set. The standard deviation for our five points is 42, an unusually
large value caused by the outlier data point of 100. A
robust alternative to the standard deviation is something called MAD: the median absolute deviation from
the median. In our case the deviations (in absolute
value) of the data 7,8,6,4,100 from the median 7 are
0,1,1,3,93; the median of these deviations is 1, so the
MAD of our data is 1--a far cry from the standard deviation of 42.

Exploratory data analysis also employs a variety
of visual and graphical devices to summarize data. One
example of such a device is the box plot invented by
John Tukey, one of the pioneers in exploratory data
analysis. A box plot consists of a rectangle around
the middle 50% of the data, to emphasize the location
of the majority of the data, together with straight
line "whiskers" to represent the range of the upper and
lower quartile. For our data the box would look something like this:

Whereas traditional statistics is primarily concerned with formal inference and estimation of

parameters, data analysis explores data in a wide
variety of ways to gain insight, to summarize information, and to expose unanticipated phenomena. Data
analysis uses many different methods, especially graphical methods, to summarize and explore; it emphasizes
methods that are robust, versatile, and insensitive to
errors of measurement or unfulfilled assumptions.
Graphical methods display data as a photograph portrays
life; robust methods are like insurance policies: they
protect against the potentially catastrophic effects of
small amounts of bad data.

　　　Exploratory data analysis can put the excitement
of discovery back into the teaching of statistics. A
few "aha" experiences can make the subject far more
exciting than grinding through formulas and significance levels. Moreover, as arithmetic provides a concrete base for the study of algebra, so exploratory
data analysis provides valuable experience in manipu-
lating real data before encountering the more abstract
aspects of classical statistical inference. Much of
exploratory data analysis requires only arithmetic and
graphing, so it is easy to introduce it at many different points in the school curriculum.

References

[1] D.R. McNeil. Interactive Data Analysis: A Practi-
     cal Primer. Wiley, 1980.
[2] John W. Tukey. Exploratory Data Analysis.
     Addison Wesley, 1977.
[3] P.F. Velleman and D.C. Hoaglin. Applications,
     Basics and Computing of Exploratory Data Analysis.
     Duxbury Press, 1980.
[4] R. McGill, J.W. Tukey and W.A. Larsen. "Varia-
     tions of box plots." American Statistician 32
     (1978) 12-16.

Round and Round at the Round Table

Several roundtable discussions at ICME IV provided interesting commentary on issues of current importance in mathematics education. Here are some excerpts from these discussions:

On Problem Solving

The development of problem-solving ability should direct the efforts of mathematics educators through the next decade. Performance in problem solving will measure the effectiveness of our personal and national possession of mathematical competence.

-- NCTM, An Agenda for Action

Problem solving. "What does that mean? What is mathematics other than problem solving?"

-- Douglas Hofstadter

Problem solving is learning to do something intelligent in a new situation, in a new field.

--Henry Pollak

I distinguish between mathematical activities that involve reasoning and mathematical activities that are carried out more or less automatically. The latter are

learned by a great deal of drill and practice, but do
not involve any evaluation, examination, or integration
of knowledge. In contrast, problem solving is an
activity that involves reasoning in terms of operations
on data in order to come up with something that is not
an immediate and direct consequence of the data.

-- Robert Karplus

   Problem solving involves applying mathematics to
the real world, serving the theory and practice of
current and emerging sciences, and resolving issues
that extend the frontiers of the mathematical sciences
themselves.

-- NCTM, An Agenda for Action

   Of course, the problem solver must have the "basic
facts" at his disposal. But recent work in cognitive
science stresses the difference between factual
knowledge and procedural knowledge. The latter
includes a knowledge of the conditions under which a
particular procedure may or may not be "legal," to what
arguments it applies, and so on. ...In addition to
"knowing" something (that is, being able to discuss it
when asked about it) one must know when it is relevant
to a particular problem.

-- Alan H. Schoenfeld

   In my opinion, nobody can ever learn to be a
better mathematician from reading Pólya's volumes.
You'll know more math after reading them, but they are
not going to turn you from a class B to a class A
mathematician. You won't suddenly have great, deep
insights that you didn't already have. Nothing is
going to teach anybody to have great, deep insights.
There's just no way you can do it. You either have
them or you don't. That's my view.

-- Douglas Hofstadter

   Usually we think of problem solving as a higher
order activity, especially as compared to mastery of
specific algorithms. Yet frequently it is the weaker

student--the one who has not learned the algorithm--who engages in true problem solving because the only way he can solve the problem is to reason his way through it.

-- Henry Pollak

There has been little conclusive evidence to date that heuristics 'work'--in the sense that students can learn to use them, and improve their problem-solving performance thereby. In fact, for the most part, heuristics are ignored, dismissed or disdained outside the math-ed community.

-- Alan H. Schoenfeld

What I hear is people desperately trying to find a way to fight off or subvert "back to basics." One thing to do is to think about what one means by basics. What good is it to be able to multiply 7.6 times 4.7 in 8.3 seconds when you don't know whether you should be multiplying or adding those numbers? The reason the problem-solving bandwagon exists is that it is fighting off the back-to-basics movement.

-- Henry Pollak

On Teaching

If teachers would only encourage guessing. I remember so many of my math teachers telling me that if you guess, it shows that you don't know. But in fact there is no way to really proceed in mathematics without guessing. You have to guess! You have to have intuitive judgment as to the way it might go. But then you must be willing to check your guess. You have to know that simply thinking that it may be right doesn't make it right.

-- Leon Henkin

There's something that teachers do on the whole that kills children.

-- Hermina Sinclair

My child's own education is being destroyed by
going to school.

> -- A. Geoffrey Howson

The setting of expectations (with assessments)
kills teaching. We must liberate teachers!

> -- Ubiratan D'Ambrosio

You can only liberate those who are strong enough
to be free.

> -- A. Geoffrey Howson

One of the big misapprehensions about mathematics
that we perpetrate in our classrooms is that the
teacher always seems to know the answer to any problem
that is discussed. This gives students the idea that
there is a book somewhere with all the right answers to
all of the interesting questions, and that teachers
know those answers. And if one could get hold of the
book, one would have everything settled. That's so
unlike the true nature of mathematics.

> -- Leon Henkin

On Minimal Competency

One of the dirtiest expressions in this country
(U.S.) is "minimal competency." In one of our states
the student masters 1/2 in grade 1, 1/3 in grade 2, and
1/4 in grade 3.

A minimum competency item seems to be one in which
you guarantee that only 10% of the students will fail
that item. And that's because the legislators have the
money to cope with the lowest 10%. It's not minimum
for anything, and it's not competency!

Much of the problem with the minimal competency
movement is political. Those who legislate minimal
competency testing often do not know what mathematics
is beyond addition and multiplication. The biggest

mistake we made in the U.S. was to <u>not</u> first give
minimum competency tests to legislators and those who
mandate them before giving them to students.

-- Ann McAloon,
   Educational Testing Service

On the Role of Applications

You really haven't had a mathematics education
if you don't know how to use it.

-- Henry Pollak

To teach arithmetic without reference to quantitative measures of time and money, geometry without
reference to machinery and the solar system, calculus
without reference to problems of optimization, would be
to withhold from our students the very things which
give our subject its justification and meaning. There
is a sense in which the whole of school mathematics is
"applicable" or "for application," even if not
"applied."

-- Douglas Quadling

The whole practical value of education is being
questioned. Mathematics, like everything else, must
sell itself as a valuable part of education for everybody, not merely as an intellectual recreation for a
small number of highly gifted people.

-- Robert Karplus

Insofar as one emphasizes applications as a
creative aspect of modelling, one actually increases
the mathematical know how of the students. They get a
better, more diversified education in mathematics. You
not only have the student on your side--that being such
an important part of learning--so more learning is possible, but in addition there is an entirely new com-
ponent added to the student's education.

-- Samuel Goldberg

People are calling for reasoning, and applications seem to offer an opportunity for a curriculum emphasis in this direction for two reasons: pure mathematics is something teachers are accustomed to do either in drill and practice applications, or not to do at all. Very few teachers are familiar with any parts of pure mathematics that are not normally taught in a drill and practice routine. Teachers look for problem solving in an area where arithmetic and simple algebra can be used: that suggests elementary applications.

-- Robert Karplus

In every area of pure or applied science mathematics offers an important reserve of power. The implication of this for education is that mathematics should enter as far as possible in a form where it is integrated into the main subject matter; the role of the mathematical community lies more in the provision of suitable resources to support such courses than in setting up separate courses of instruction in mathematics.

-- Douglas Quadling

On Teaching Applications of Mathematics

Teaching mathematical modelling is orders of magnitude more difficult than what we have done up to now. Many teachers have never done this themselves. It is just tough. You need a much freer classroom at all stages. To the extent that you need a freer classrooom, a less prescribed classroom, you need better teachers, more secure teachers, more confident teachers. Therefore you have the odds against you in terms of the teaching that needs to be done.

-- Samuel Goldberg

It is very difficult to see how the education establishment in this country would ever permit the development of a truly interdisciplinary course: departmental structure and textbooks prevent this.

-- Henry Pollak

There are even cases of physics and mathematics
teachers in the same skin--because of the tremendous
shortage of qualified teachers. But despite this, they
do not talk to each other. They are expected to produce graduates who have proceeded through umpteen
chapters of this book or another one. To rethink what
you do so that it meets the expectations of this
environment and at the same time permits expression of
some of this overlap of ideas--that's pretty rough.

-- Robert Karplus

I do not think any approach to the teaching of
applications in the schools will be successful without
a program of inservice education. This will cost
money. The advocates of the applications have a
responsibility to wrestle with these practical problems
of implementing their objective in the real world.

-- Paul C. Rosenbloom

On Geometry

The most important geometric theorems in the
curriculum are those that deal with length, area,
volume, and similar figures. They are important
because of their biological and structural consequences. Yet very few classes ever get to those
theorems and these important connections are rarely, if
ever, mentioned in geometry textbooks.

-- Zalman Usiskin

The U.S. is the most backward country in the world
in the teaching of secondary school geometry.

-- Jean Dieudonné

I'm becoming allergic to teaching geometry courses
just for teachers.

-- Branko Grünbaum

Historical Perspective

Leon Henkin

The idea of holding international congresses each
four years is an old one for mathematics researchers.
We have been doing it since the beginning of the century.   In fact an International Congress that was held
in Oslo in 1900 had an enormous impact throughout the
century. The leaders invited the towering figure of
David Hilbert, a German mathematician, to list out-
standing problems in mathematics, and mathematicians
have been working on them ever since.

These mathematics congresses have several sec-
tions: there is a section for geometry, a section for
algebra, a section for analysis and even a section for
mathematics education. But the mathematics education
section is very much a step-child. The fact is that
most research mathematicians are not really very
interested in mathematics education.

But in the mid-1950's, for one reason or another
that need not concern us here, well-known research
mathematicians in the United States began to turn their
attention to the question of mathematics education.
They begin to look in particular into the kind of
mathematics that was being taught in the schools. What
they saw was that the mathematics being taught there
was very different from the mathematics as they knew it
and seemed to be falling farther and farther behind.

Of course in the United States there also was tremendous governmental interest, originally due to Sputnik and the realization that all sorts of techno- logical and military questions depend on mathematics. Just as these mathematicians became interested in edu- cation, so did the government. The two came together, and many millions of dollars were provided through such agencies as the National Science Foundation. These mathematicians began by suggesting changes in the material that was taught; they prepared new books and entire new curricula. Since America was viewed as a leader, these ideas spread to Western Europe and subse- quently to developing countries, where they are still going on.

Among the mathematicians who turned increasing attention toward education was Hans Freudenthal of Hol- land, one of the four principal speakers in this Congress. He was largely instrumental in persuading his colleagues that a single section in the Interna- tional Mathematical Congress is not really enough for mathematics education. It doesn't give adequate scope to bring new ideas forward. Among other reasons, those few mathematicians who began to go into mathematics education soon found that to proceed in a satisfactory way, they had to join forces with people who were not mathematicians--linguists, for example, and psycholo- gists. None of these several groups who had been work- ing separately had ever gotten together to talk with each other. Since these other groups would feel com- pletely out of place in the Mathematics Congress, they decided to have their own kind of congress, an Interna- tional Congress of Mathematics Education.

The first one was held in Lyon, France, in 1969. It was a rather small affair. Perhaps the most impor- tant thing that emerged was the impulse to proceed on a regular basis. The Second Congress was held only three years later, in 1972, but from then on it was declared that these Congresses would be held every four years, alternating with the International Congresses of Mathematics. The Second Congress was held in Exeter, England, the Third in Kahlsruhe, Germany, in 1976 and now the Fourth Congress is here in Berkeley in 1980. These Congresses are very much interdisciplinary

meetings in which people whose work overlaps but who
normally don't have close working relationships have a
chance to exchange ideas, to find out about things that
may have appeared remote and suddenly are seen to be
closely related.

The program for ICME IV was put together by an
international committee chaired by Henry Pollak of Bell
Labs. Several meetings were called in which people
from Australia, from Germany, from Africa, and from
other distant places, came together for a meeting of
about three days to hammer out what the program should
be. Literally thousands of letters went out to mathematicians, psychologists, administrators, classroom
teachers, to all sorts of people who had an interest in
mathematics education, asking: "What do you consider
the important questions in mathematics education? Who
are the people in your country who are doing exciting
and interesting things in that area?"

In this way a large list of names and subject
areas was brought together, so the committee was able
to identify certain topics of world-wide interest. The
Committee then formed program items on these topics, so
for example, a group of four different speakers from
widely scattered regions would be asked to speak on a
common subject. That's something which distinguishes
this Congress from any I had seen before. The usual
custom in the academic world when you invite a speaker
is to ask him what he would like to talk about. It was
not done that way in this Congress and I think the
result has been very fruitful.

Reports

Geometry under Siege

Donald J. Albers

"Euclid Must Go!" With these words Jean Dieudonné
attacked the standard one-year course in Euclidean
geometry at a 1959 international conference devoted to
improving mathematical education. Naturally his com-
ments provoked a great controversy. His remarks at
that meeting help to explain the battle that followed:

> Let us assume for the sake of argument that
> one had to teach plane Euclidean geometry to
> mature minds from another world who had never
> heard of it... Then the whole course might,
> I think, be tackled in two or three
> hours...one of them being occupied by the
> description of the axiom system, one by its
> useful consequences and possibly a third one
> by a few mildly interesting exercises.

> Everything else which now fills volumes of
> "elementary geometry"...and by that I mean,
> for instance, everything about triangles (it
> is perfectly feasible and desirable to
> describe the whole theory without even defin-
> ing a triangle!) almost everything about
> inversion, systems of circles, conics,
> etc.,...has just as much relevance to what
> mathematicians (pure and applied) are doing
> today as magic squares or chess problems! [1]

Twenty-one years later we find Dieudonné at ICME
IV participating in sessions on "The Death of Geometry
at the Post-Secondary Level." The passage of time has
been marked by several attempts to reform the geometry
curriculum, resulting in new teaching materials from
various curriculum groups. These materials approach
geometry from such diverse perspectives as transformations, coordinates, vectors, affinities, and linear
algebra. In most countries, it is now difficult to
find the traditional year-long course in Euclidean
geometry, but in the United States Euclid remains
firmly entrenched as a standard one-year course.

The health of geometry at all levels was the subject of vigorous discussions at ICME IV. Zalman Usiskin of the University of Chicago, co-author of a high
school level transformational approach to geometry, is
not optimistic. "If you take out the [present]
geometry from the high school, it doesn't hurt a thing.
The only thing that has kept geometry as a one-year
course in many states of this country [U.S.] is the
fact that certain colleges still require geometry for
admission and many guidance counselors still think that
most colleges require it for admission even though most
of them don't."

Both Dieudonné and Usiskin suggest that the classical, rigorous, one-year course in Euclidean geometry
has little to do with additional study in mathematics
or its applications. "We are in trouble," said Usiskin.
"If you look at the geometry done by mathematicians, it
is a geometry of coordinates, or a geometry of
transformations, or a geometry of convexity, or a fin-
ite geometry. If you look at those four things and
then you look at the geometry we teach, the intersec-
tion is nearly empty. The present content is not useful in later mathematics or in relating mathematics to
the physical world; it has been selected simply because
it is easily proved from a particular set of proposi-
tions."

Joe Crosswhite, professor of mathematics education
at Ohio State University and the author of a high
school geometry text, participated in an exciting round
table discussion at ICME IV on the decline of geometry.

Crosswhite spent eight years teaching at the secondary
level before moving to Ohio State, and he underscores
Usiskin's last point: "Most mathematics teachers in the
United States have never claimed to teach geometry for
geometric content itself, but for process objectives,
that is, for the process of deduction and axiomatiza-
tion."

Crosswhite's statement echoes a commonly cited
reason for teaching geometry: instilling the deductive
process in the minds of students. Over the years, this
general purpose of a rigorous geometry course has been
expressed in various ways: to develop critical thinking, or to give life reasoning, or to improve the mind.
"You can see the steps and get the idea..." said Robert
Osserman of Stanford University, speaking of many people for whom geometry was a revelation. "I don't know
if this [deductive approach] is the best way to develop
in students the ability to see what a really rigorous
argument is, but it works. It has worked!"

According to Usiskin, however, the best educational research shows that the standard geometry course
does not help to develop critical thinking [2]. This
research confirms the opinions of many university
mathematicians and curriculum reformers, but it con-
trasts sharply with the results of a recent (1975) survey [3] by George Gearhart of high school geometry
teachers. "Emerging from the survey is a picture of
the role teachers believe geometry should play in the
secondary curriculum: The importance of geometry was
reaffirmed, and teachers are apparently satisfied with
the usual one-year course devoted to the subject."

One clear point of agreement with high school
teachers emerged from the Berkeley discussions about
geometry: Geometry is important. "We teach geometry
because it is a carrier of culture," claims Seymour
Papert. Branko Grünbaum of the University of Washington agrees, but expresses his view in a different way:
"Geometry is one of the few approaches our senses give
us to the outside world."

"To do mathematics," said Jean Dieudonné, "you
need fundamental geometric insight. Ninety percent of

modern mathematics is permeated with geometric
insight." "Most important to me," added Hans Zeitler of
Germany, "is that we need geometry at school. The mental development of a child starts with visual perceptions. Every child wishes to and also has to grasp its
visual environment. Children paint. Children draw.
They are by nature little geometers."

Elementary School Geometry

Each of these reasons speaks strongly to the
importance of geometry in the elementary school. Participants at the Congress were slightly more encouraged
about the state of geometry in elementary schools than
about geometry in secondary schools. "I am more
optimistic about elementary school geometry than I am
about secondary school geometry," commented Usiskin.
"I see elementary school geometry as getting more wide
ranging and more intuitive." Crosswhite agrees, momen-
tarily lifting one's hopes before dashing them: "If you
look at elementary school textbooks, geometry is very
much alive. The geometric content is better than ever.
But in many cases it's not being taught. Pressures in
this country are very much on algorithmic things. So
geometry chapters are the most easily skipped in school
books through grade 8!"

Crosswhite described recent results of the
mathematics portion of the National Assessment of Educational Progress. "Our N.A.E.P. data on 8th graders
shows that what students can do is name common
geometric figures, but they cannot use any properties
which characterize figures." According to Crosswhite,
this clearly means difficulties for students who are to
enter the standard one-year geometry course and points
out another reason why the present geometry course may
be doomed to failure. "The conceptual structure
(congruence, similarity, etc.) of geometry is not built
by the time they are age 15. Then they are placed in
the position of building a deductive system out of concepts with which they are not familiar."

Ann McAloon of the Educational Testing Service
cited test data which also suggests a decline in
geometry. "On the General Educational Test for high

school students, any test item involving a geometric
figure prompts an 'omit'." She fears that the situation is worsening, citing trends in the Graduate Record
Exam and the National Teacher Exam: "In the GRE and the
NTE, there is an increasing number of omits, especially
on geometry items."

Branko Grünbaum and Robert Osserman were not
surprised by this data. "There is abroad in this land
(and Europe not excepted) a great amount of geometric
illiteracy," said Grünbaum. "Things and facts that one
would normally assume people are aware of, such as simple spatial relations, are completely absent from their
knowledge." Robert Osserman concurred: "Although
geometry has permeated almost all of mathematics, it
has been neglected as a subject in its own right."

Must Bourbaki Go?

Osserman went on to identify the Bourbaki approach
to mathematics as the villain in the decline of
geometry. "The neglect of geometry is due in the main
to the Bourbaki approach to mathematics, which asserts
the primary of the general over the particular, the
concrete, and the applied. What I fear is that much of
geometry falls into that concrete, unstructured part of
mathematics. In particular, this is where the notion
of geometric intuition reigns supreme." In support of
this charge he offered the opinion of another geometer,
Michael Spivak, who was not present at the Congress:

> Bourbaki is the originator of that famous
> pedagogical method whereby one begins with
> the general and proceeds to the particular
> only after the student is too confused to
> understand even that anymore. [4; p. 610]

Osserman sees a curious feature in the highly structured Bourbaki approach. "As you follow this approach
down the path, you often find at the end of the road,
as did Gertrude Stein in Oakland, that 'There is no
there there.' It just seems that the original subject
of study has vanished in its pursuit."

Certainly the influence of Bourbaki has been felt
in U.S. mathematics education. Much of the "new math"
can be traced to Bourbaki's influence. According to
Crosswhite, U.S. mathematics teachers become preoccu-
pied with generalization of axioms in the 1950's and
1960's when mathematicians began to get excited about
doing modern mathematics in the schools. "They advo-
cated an axiomatic, abstract approach. And the teachers misunderstood in many cases, and tried to do for
the run-of-the-mill 10th grade [geometry] student what
they would do for the future mathematician."

Listening to these charges, one might get the
impression that Dieudonné's 1959 assertion that "Euclid
Must Go!" is now being replaced by "Bourbaki Must Go!"
Dieudonné, a long-time member of Bourbaki, responded
vigorously to these attacks. "I claim that general and
abstract ideas are used because they work! Complaints
about the abstract approach are not a new thing. It is
just sour grapes! We [Bourbaki] are very much against
abstraction for its own sake. Bourbaki was not
intended as a pedagogical venture. It was intended for
people who are doing research at the Ph.D. level. If
some ill-advised people thought Bourbaki could be used
for pedagogical purposes at lower levels, it's their
fault and not ours."

Solutions

What's to be done? How is the decline to be
reversed? All of the round table debaters agreed that
colleges and universities can play a pivotal role in
revitalizing geometry in the schools. "The future is
not good for [school] geometry," summarized Usiskin,
"unless the colleges start teaching geometry again."

Osserman believes that university mathematicians
are already seeing signs, at the research level, of the
mathematical pendulum swinging away from the highly
structured Bourbaki approach and back to one in which
geometry will play a bigger role. He thinks that much
can be done to facilitate the filtering-down process
from mathematical research levels to the classroom.
"Universities should institute, reinstitute, or recon-
stitute courses in which the fragile but vital notion

of geometric intuition is fostered and nurtured."

Grünbaum agrees, arguing that colleges and universities should place a greater emphasis on intuitive
geometry. "The visual aspects of geometry in colleges
in this country are practically non-existent. We must
do much more at the university and college levels. We
cannot expect high school teachers to suddenly have a
revelation [on intuitive geometry] and to do something
for which they have not been prepared."

Turtle Geometry

One radically different solution is a new device called Turtle Geometry, a computer system
developed by Seymour Papert. "My thesis about what's
happening is that for various historical reasons we try
to impose on children a kind of geometry which is out
of step with their intuitions. We have learned from
experience that children are unable to grasp the conceptual basis and relate to that geometry. We have
falsely concluded that because in this geometry the
children are unable to grasp the concepts, they are
unable to grasp new concepts in all geometries."

In Papert's lab children type in programs that
control a simple robot called Turtle, making it travel
intricate and interesting paths. They write their programs with an elementary computer language called LOGO.
(The robot Turtle can be replaced with an image moving
on a TV screen in which the computer draws a picture of
the path that the Turtle would have taken.)

"There is a tremendous fascination for children
with visual effects, computer games, 'Star Wars'. When
we can create computer systems where the children are
able to write programs to create these sorts of effects
themselves, moving objects around on a screen, then
they become deeply and passionately involved.

"After a while they start posing questions like:
'How can we draw a circle?' In Turtle Geometry classes
we don't give them an answer. We tell the child: 'Look
at yourself!' We point them to the most powerful and
richest source of mathematical knowledge there ever

was: themselves, their own body knowledge, their own
intuitive geometry. Since the time they were babies,
they were learning to get around in space.

"Then we say: 'Make a circle. Walk in a circle
and describe it.' After a while, they see: "Forward a
little, turn a little; forward a little, turn a little;
forward a little, turn a little" makes a circle. (You
might say it really makes a polygon, but that's another
question...) Once the child gets his Turtle to move in
a circle, you've won. That child is mathematized."

In the opening plenary address at ICME IV, Hans
Freudenthal asked whether we can teach geometry by having the learner reflect on his spatial intuitions.
Papert claims that Turtle Geometry provides an affirma-
tive answer to Freudenthal's question. "Turtle
Geometry is a computational geometry, a process
geometry, a dynamic geometry, a differential geometry.
It can put the child in touch with the kind of
mathematics which lends itself fantastically well to
the process of observing your own physical activities,
your own intuitive geometric knowledge, your own body
geometry, and using that as the source of geometric
knowledge. Turtle Geometry is the best way into
geometry, the best way into mathematics. And I say
this is what will save geometry from the death some
people are complaining about at this conference."

References

[1]   Fehr, Howard F. (Ed.) New Thinking in School
      Mathematics: Organization for Economic Cooperation
      and Development. Report of the Royaumont Seminar,
      Paris, OECD, 1960.
[2]   Fawcett, Harold P. The Nature of Proof. Thir-
      teenth Yearbook of the National Council of Teach-
      ers of Mathematics. Columbia University, 1938
[3]   Gearhart, George. "What do mathematics teachers
      think about the high school geometry controversy?"
      Mathematics Teacher 68 (1975) 486-493.
[4]   Spivak,, Michael. A Comprehensive Introduction to
      Differential Geometry, Vol. 5. Boston: Publish or
      Perish Press, 1975.

Language and Mathematics

Anthony Barcellos

With singular success children in every part of
the world master the essentials of an extremely compli-
cated body of information: their native spoken language. This remarkable feat of learning is attended by
none of the trappings of formal education. There is no
syllabus, text, classroom, or instructor. Spoken
language skills are acquired almost as a by-product of
the mundane activities of everyday life.

In his plenary address at ICME IV, Seymour Papert
asked whether mathematical skills could be acquired in
a similar manner. Can the environment of a child be so
structured as to instill naturally the essentials of
mathematics? He answers his own question affirmatively
and offers a vision of the child in dialogue with a
computer as the analog of the child in conversation
with adults:

> Computers are the Proteus of machines: they
> take on many different forms. One of their
> manifestations is as mathematics-speaking
> beings. If children grew up surrounded by
> such beings, the learning of mathematics
> might very well be much like the learning of
> spoken language. [1, p. 232]

Computers afford students opportunities to inter-
act with mathematics. Geometric objects may be brought

to life and manipulated on video displays. The
trial-and-error, give-and-take approach by which spoken
language is perfected is emulated in the exchanges
between student and computer. However, the requirements of computer programming languages are a formidable obstacle in the way of Papert's vision. He and
his research group at MIT seek to provide access to
computers for children who have yet to master written
communication. Their working motto is "No threshhold
and no ceiling."

Children as Experimentalists

Children are natural and apt practitioners of the
scientific method. As Piaget has shown, they frame
hypotheses and modify them as required by contrary
observations in the best tradition of experimental science. These hypotheses are devised by the children to
explain to their own satisfaction aspects of their
environment which they do not understand.

Young children of no more than two years of age
have been observed by Hermina Sinclair and her colleagues as the children play with tubes, rods, and
other simple objects. Says Sinclair, "I've seen all
our subjects doing the following thing: They pick up a
tube and, because putting things into things is a very
interesting activity for babies, they pick up a stick
and try to put it into the tube. So it goes! It falls
out, and they're surprised. What do they do now? They
immediately reproduce the event: carefully put it in
and already look to see where it's going to come out.

"Now these are the ingredients of experimentation,
and they're there at the age of 16, 17, 18 months.
That's the kind of thing Seymour (Papert) was also
talking about: You have an image of what you want to
happen and either it comes or it doesn't."

According to Sinclair, current elementary instruction in mathematics fails to consider that children
already have ideas and theories about numbers. Children notice numbers on signs, on labels, on houses, on
everything about them. They do not stifle their specu-
lations while awaiting revealed wisdom from adults.

Sinclair and her collaborators have established that
very young children believe that numbers and words
should have certain special relationships. Given a
picture of three identical objects and asked to label
the picture, children demonstrate that they feel each
object should be provided with a "squiggle"--a label--
of its own, exhibiting a strong sense of one-to-one
correspondence.

Similarly, children expect words describing large
objects to be correspondingly large. Sinclair cites
two words which many young children would consider
inappropriate for the objects that they name:
"elephant" and "butterfly." The one seems too small
for the object it names, the other too large.

This propensity to imbue language with quantitative aspects has not been exploited in the educational
process. Yet this indication that children generate
their own primitive notions of mathematics suggests
that an environment which stimulated this trait could
advance mathematical learning in a manner similar to
the language-learning process.

It is in this context that Papert sees the computer as the versatile instrument which permits the
child to deal directly with mathematics, much as in
daily life he has recourse to speech.

"Also what Seymour was saying," remarks Sinclair,
"was that children are fascinated by these funny things
that are obviously important to adults. And with this
I quite agree. But there's a difference between being
fascinated--and understanding that this is powerful for
adults--and this step where you know that you can use
the stuff yourself. You use a letter and something
happens. Everybody's noticed how mad children are
about the telephone. You have to keep them away from
those numbers; otherwise they'd be ringing up the
weather bureau in Tokyo or something. But why? Because you use numbers and something happens. And
that's the case with those computers. You use symbols
--never mind what--and something happens. Until now
there was for children just an annoying feeling that
squiggles for adults must make things happen, because

otherwise--well, why would they be so busy with them
all the time? But this computer stuff is perfect. Not
only can you make something happen, but you can change
it in subtle ways. Now whether you call that language
or not, it is the basic stuff of language."

Given the opportunity to control with very simple
commands the motion of an object on a computer screen,
children readily enter into investigations of the paths
which they are able to trace. Using the LOGO system
developed by Papert at MIT, children quickly master
rectilinear motion. As the children become more ambitious they tend to trace out curved paths, perhaps cir-
cular. They soon re-invent polygonal approximation.
Papert declares that preschool children can achieve
through continued interaction with computers an intuitive grasp of dynamic concepts of geometry denied those
limited to crayons and paper.

Papert's ideal is each child with his own computer. The computer is to be regarded as an instrument
as basic as the pencil, rather than as an exotic device
introduced into the educational process.

The computer affords the child greater freedom to
experiment with mathematics than such purportedly flex-
ible education programs as the "discovery" method,
which, in practice, may result in students being herded
toward a goal rather than led. Unlike the sterile
question-answer format of present computer-assisted
instruction--in which the computer is programming the
child, as Papert puts it--the computer will not sit in
judgment on the child.

"If you find your theory just pushed away by
authority," says Sinclair, "your inclination to make
theories to master your environment is going to drop.
Just imagine what would have happened to physics if
physicists had let go of their theories with the first
unfortunate experiment. If you have a theory, and you
let it go straightaway, you never get anywhere. And
this is the same for children. The whole human attitude of one who wants to master his environment is to
set up a theory and hang on to it until one sees that
there's a better theory. If you don't hang on to your

theories, the world crumbles into objects that behave in the queerest ways. One of the bad messages that's given to small children is, 'Look, forget it. Your theories are not worth it. Do as I say.' I'd like to see teachers very much more aware both of history and of children, to be able to see this."

Learning Mathematics in Mathland

Once the computer is accepted as the proper tool to stimulate children's inclination to experiment and to learn, Papert sees a fusion of mathematics and language as a generalized form of communication:

> We are learning how to make computers with which children love to communicate. When this communication occurs, children learn mathematics as a living language. Moreover, mathematical communication and alphabetic communication are thereby both transformed from the alien and therefore difficult things they are for most children into natural and therefore easy ones. The idea of 'talking mathematics' to a computer can be generalized to a view of learning mathematics in 'Mathland;' that is to say, in a context which is to learning mathematics what living in France is to learning French. [2, p. 6]

In this Papert sees the answer to the question he poses as the fundamental problem of mathematics education: "Why is it necessary to teach children mathematics?" Clearly, Papert expects the computer to reform mathematics education to the point of abolishing it.

References

[1] Seymour Papert. "Education forum: New cultures from new techniques." Byte 5 (Sept. 1980) 230-240.
[2] Seymour Papert. Mindstorms: Children, Computers and Powerful Ideas. Basic Books, 1980.

Computers in the Classroom

Lynn Arthur Steen

Mathematics by tradition has been a subject
requiring only the simplest of tools. For Archimedes,
it was sufficient to have a stick and some sand; for
Fermat, larger margins in his books might have helped.
Even for modern research mathematicians, the normal
tools are a bookcase, a blackboard, and a large waste
basket.

But modern technology has changed all that.
Mathematics classrooms in many parts of the world are
now equipped with hand calculators. (Even if the
teachers don't have them, many students surely do.) In
recent years the new small personal computers have also
become visible in some school classrooms, and many colleges have even begun development of computer labs
associated with traditional mathematics courses. At
the research level, some areas of mathematics have
become so closely linked to computer methods that
mathematics is now partly an empirical undertaking.

These changes in mathematics and mathematics education were reflected at ICME IV by numerous sessions
on the role of computers and calculators in the classroom, on the new mathematics associated with computer
methods, and in a plenary address by Seymour Papert on
the computer as a "carrier" of mathematical culture.
Although delegates from different countries approached
these issues from quite different perspectives, there

seemed little dispute over the central theme--that computers are ushering in a totally different era for mathematics education.

Mathematics and Machines

Mathematics is, to a great extent, a man-made universe. The computer is a machine that can actually create a universe under the direction of man. Seymour Papert, in his new book Mindstorms [1], describes the computer as the Proteus of machines. "It can take on a thousand forms and can serve a thousand functions; it can appeal to a thousand tastes." Papert argues forcefully that a computer, when properly used, can enable even a very young child to begin developing the habits of mind that form the basis for mathematical thought.

Physicist Robert Karplus, newly appointed Dean of the School of Education at the University of California at Berkeley, echoes these sentiments: "Education must provide for spontaneous learning by students, as well as for instruction in previous cultural experience. I see the computer as affording the student opportunities for spontaneous or autonomous activities, while at the same time transmitting some of our cultural traditions, but in very subtle ways.

"The great opportunity of the computer is that it provides a way of dealing with students individually to enhance their sense of self-confidence and their ability to act, while at the same time providing experience with the rest of the culture in a gradual, neutral and individually tailored fashion."

Much the same things have been said in the past for previous technological innovations, notably television, and calculators. They would, according to their advocates, provide a means for individualizing instruction; they would enable students to relate better the world of the classroom with the world of work. Now, of course, both television and calculators are often blamed for many of the ills of modern education. Children who get hooked on watching television and using calculators, critics argue, never learn to read or to calculate: they grow up with neither the motivation

nor the discipline to master these important but difficult childhood skills. Even the advocates of educational television and classroom calculators generally
admit that these technological innovations have made
very little significant positive impact in the
mathematics classroom.

"The most important thing that we could do for
elementary school teachers," remarked Henry Pollak, "is
to get them to think of calculators as useful aids and
partners in their teaching, rather than as instruments
of the devil. Particularly at the elementary level,
calculators are considered to be just plain immoral.
Once you have calculators, you can afford to get data
and problems from the real world. You are no longer
frustrated if the numbers don't happen to come out
easy."

The parallels between television, calculators and
computers are sufficiently striking as to bear repeat-
ing. Three decades ago television sets were expensive
and only the rich could afford them; now in the industrial world almost every home has a television set, and
they are common in most of the third world as well.
Hand-held calculators followed the same progression two
decades later, moving from an expensive novelty to a
common (almost throw-away) household gadget in the
space of just ten years. Moreover, the labor-intensive
nature of assembly of these electronic devices makes
production relatively easy, even in the less industri-
alized countries.

Risks and Opportunities

Small computers in 1980 are approximately where
hand calculators were in 1970: expensive novelties for
the household, but essential machines for business. If
one argues only by analogy, one might expect that by
1990 personal computers would be as pervasive in our
society as calculators and televisions are today. If
this is the case, the mathematics classroom of 1990 may
bear little resemblence to the mathematics classroom of
today. The real question is this: what impact would
such computers have on mathematics education?

Many delegates at ICME IV seemed inclined to agree
with the general thrust of Papert's address, that the
introduction of computers as a common accouterment of
modern society will have a near-revolutionary effect on
mathematics education. The strains imposed by the computer are already becoming evident in the United
States--where home computers are more widespread than
elsewhere--and give hints of what is to come:

-- Increasing numbers of students now learn how to
   program computers so early that they are misfits
   in the standard mathematics curriculum.

-- Since most teachers know very little about computers, they frequently must teach children who
   know more than they do, an unfamiliar and unset-
   tling experience.

-- Demand for computer-related mathematics (algorithms, discrete structures, combinatorial
   analysis) is increasing, even while school systems are trying to cope with traditionalist
   demands for "basics."

-- Business demand for individuals trained in computers is so high that very few students are
   eager or even willing to prepare for careers in
   mathematics teaching.

These are the superficial issues surrounding the
role of computers in mathematics education: they rest
near the surface, and are easy for any serious observer
to recognize. This does not mean that they are easy to
solve, however. But there are deeper problems as well,
relating not just to the presence of computers but to
their nature. Papert cites two of these problems as of
major importance:

"Education has a very bad track record in the use
of technology because it tries to do with the new thing
exactly what it was doing before. This is what has
happened with the computer. The easy way to use the
computer is to mechanize those aspects of the instruction process that are most easily formalized, like
drill and practice. The really serious danger is the

professionalization of people involved with computers
in education, people highly focussed on that particular
use of the computer. Unfortunately, it is not just a
technological bias; it is, rather, highly ideological,
related to specific viewpoints on the role of students
and teachers in the classroom.

"There has been a balance of sorts between the
neo-behaviorists and the advocates of open education,
at least in the United States. The primitive uses of
the computer lend themselves much more easily to the
neo-behaviorist trend, which is therefore likely to be
enormously strengthened. In fact, conferences on computers in education today are totally dominated by
thinking that comes from that sort of intellectual
tradition.

"The other major danger of computers are social.
Computers are made by engineers for engineers: they
express a certain set of values of what is easy, what
is important, and what is good to do. These values
resonate with a class of people that is very white,
very male, and very technologically oriented. Insofar
as the computer is effective in facilitating access to
mathematical knowledge, we are going to see more
privilege going to the already over-privileged. There
is a very serious danger that the computer, which could
help democratize education, might create a tighter,
more closed and more elitist subculture."

The dangers that Papert sees, like his vision for
the computer as a carrier of mathematical culture, are
rooted in his belief that computers will have a major
impact on education, whether for good or for ill. Others see the problems growing more gradually, and are
concerned more with potential dangers than with current
abuse. Karplus, for instance, points out that most
major computer manufacturers are now focusing on the
home market rather than on the schools: "The danger I
see is really suggested by the experience of television. Television had opportunities similar to those
that the computer has today. But the commercial
development of television created what I consider to be
a cultural tragedy in the United States. I am afraid
that the effort at commerical development of computers

will be the same, to sell things in a way that appeals
to people as they were two or three years ago when the
engineers got their conception of what people were willing to buy."

George Immerzeel of the University of Northern
Iowa underscored these concerns that decisions about
the direction of computer development will be made by a
small class of the technologically elite. It is
vitally important, commented Immerzeel, that such people be convinced that "thinking is a basic skill."

Evolution or Revolution?

Peter Hilton, Secretary of the International
Commission on Mathematical Instruction and a member of
the organizing committee for ICME IV, also expressed
concern about the educational impact of computers.
"There will be a change," predicts Hilton, "but it will
come slowly. It will come by a much more continuous
process than the word 'revolution' suggests. Papert's
vision seems to suggest that we will solve the problem
of the inadequacy of our teachers by eliminating them
as a factor in the educational process. I find that a
disquieting, even distasteful, notion."

"That children who do have access to a computer at
an early stage have a set of experiences that are quite
different from those who do not have such access to
computers I think is absolutely true. And this is
potentially very important for their development. But
I do not believe that we are going to see within the
next decade anything like the sort of revolution that
Papert's presentation suggested. If we do see it, it
will be seen in the highly developed countries. Here
again I think we must be terribly careful, because I
could imagine that Papert could give a lot of offense
to those participating in ICME IV from the developing
countries for whom these things are far from even the
glimmmerings of an actuality.

"I do not think computers will eliminate the need
for good teaching, and for concentrating on the role of
mathematics in the child's developing life. I do think
it will give us the possibility of involving the child

much more actively in quantitative and geometrical
ideas. In this respect it is clearly a very good
thing."

Hilton shares the concern of many educators that
the ability of the computer to accelerate mathematical
maturity may impose new strains on the mathematics
classroom. However, he argues that these strains are
neither unique nor revolutionary: they are just natural
extensions of problems teachers have faced for years.
"It is comparable to the difference that already exists
between the child whose parents involve themselves in a
very positive way in the child's education and the
child whose parents do not. So I think it is the sort
of problem we have already had, although perhaps magni-
fied by the computer.

"I could envisage that the computer could very
well make the child increasingly impatient with dull
presentation of dull material. If the child is really
having an exciting time at home with the computer,
thinking interesting thoughts, and then has to go to
school and do a lot of rote calculations, I can imagine
that that will turn a child off even worse than he or
she is turned off now. There is a real risk there."

Entering the Computer Age

A recent report [2] by the U.S. National Science Foundation and Department of Education chides public education (in the U.S.) for ignoring the impact of
the computer revolution. "Just as we recognize the
Stone Age and the Bronze Age, the Iron Age and the
Machine Age, historians are likely to look back on our
own time and label it the 'Computer Age.' ...Examina-
tion of school curricula, however, would, by and large,
offer little evidence of the existence of this electronic revolution."

The slow response of school curricula is probably
just a normal symptom of the natural conservatism of
educational systems. However, market forces are bringing computers into business and even into the home
ahead of the classroom; these same market forces are
pulling mathematics teachers out of the classroom and

into the high technology industries. Computer technology is spanning the globe, adding a third element to
the information-communication complex to which Ubiratan
D'Ambrosio refers in his essay. This revolution is
increasing both the potential and the tensions involved
with mathematics education. Whether computers will be
viewed as an "instrument of the devil" or as a "carrier
of mathematical culture" is one of the major challenges
facing mathematics teachers in the last two decades of
this century.

References

[1]   Seymour Papert. Mindstorms: Children, Computers,
      and Powerful Ideas. Basic Books, 1980.
[2]   Shirley M. Hufstedler and Donald N. Langenberg.
      Science and Engineering Education for the 1980s
      and Beyond. National Science Foundation, 1980.

Universal Primary Education

Anthony Barcellos

Universal primary education is a goal yet un-
achieved in many developing nations of the world. Significant portions of the populations of these nations
are entirely without formal education. Many African
countries have less than twenty years' existence as
independent states, and another twenty years may pass
before they attain universal primary education.

This topic figured prominently in the ICME IV program. Henry Pollak, Chairman of the International Program Committee for the Congress, reported that universal primary education was suggested as an ICME subject
more than anything else. "When we began to organize
the Congress, we asked around the world about the major
issues in mathematics education that educators and
mathematicians were most concerned about. The one
above any other that concerned everybody was universal
primary education. What should be the content and the
point of view of those few years? In many countries
the children only get four to six years of education.
What should you do with that time? How do you mix the
mathematics with its applications? How should it be
different from country to country?

"Many parts of the world are very concerned with
that because they are really trying for universal primary education for the first time. In the United
States universal primary education has been here for

some time, but that has not lessened our quarrels and
concerns over what should be in it."

In countries where the average educational level
is four years of schooling, the effective use of that
brief period of formal education becomes an optimization problem of crucial importance. Bienvenido Nebres
of Ateneo de Manila University asks, "What can you hope
to achieve in four grades? Instead of teaching
material which is preparatory to high school, should
you reconsider, and accelerate certain material? If
they drop out after four grades you cannot wait until
sixth grade to teach them division."

However, a narrowly prescribed solution to this
problem could give rise to another difficulty: the im-
position of an artificial limit on the education which
students may obtain. Unless provision is made for the
extension and continuation of education as the development of the society permits, the brief terminal education program could stunt the growth of its host
society.

"The problem," observes Nebres, "is that the curriculum is not set up that way at all. The textbooks
and schools are set up on the assumption that these
students will go on to high school. But one of the
problems that many developing countries face is that if
you raise this question—how can we design the curriculum so that it is terminal at fourth grade—the question always comes: Aren't you being unfair to those
who go on?"

Global Uniformity

In another sense, however, current primary education programs are already universal: "universal" in
the sense that the basic education curriculum the world
over is modeled on that of western society. Without
regard for cultural context or societal needs, the subject matter taught in the classes of the world is pat-
terned after that of Europe and North America. In particular, truncated or compressed versions of a curriculum intended to lead to advanced studies are being
offered as terminal courses.

"I would say that the primary school curriculum
throughout the world is basically the same," declares
Ed Jacobsen of UNESCO. "It's the same, you'll find, in
almost any country in the world. It certainly is true
that if you go into an African school, it's the same as
going into an English or American school. Yet the
needs of the children are certainly very different.

"The argument put forward by governments is that
mathematics and science are universal, and therefore it
doesn't matter. But what is more needed is the ability
of the people who have studied mathematics for only six
years to be able to utilize what they have learned.
Certainly the way that it's been taught to them makes
it very difficult for them really to apply it, because
they've learned it from outside of their culture.
Hence they can't see how the mathematics and science
that they learned in school relates at all to the world
in which they live."

The uniformity of curriculum in developing nations
is an enduring vestige of their days as colonies.
Adaptions of the textbooks of the western colonial
powers were the first educational material used by the
newly independent nations. In some instances, special
editions for use in the third world were written by
western educators. In general, today's texts derive
directly from these original western sources in an
unbroken line of descent.

Serving Local Needs

While the goal of providing an adequate education to each school age child depends on sufficient
material progress and development, the revision of curricula to reflect the needs of the indigenous culture
is a project for the present day. Discussions on this
subject at ICME IV identified a number of important
considerations in any program of curriculum reform.
Ideally, an indigenous curriculum should provide the
tools necessary to function in the existing society
while also facilitating the development of that
society.

"They should look at the problems that the community faces," says Jacobsen, "and how the community tries to solve these problems, and try to extract the mathematical framework--to use the mathematics to try to solve the problem.

"I've seen examples of how much mathematics goes into building an African hut, or how much mathematics exists in farming. This is what I mean by being able to extract the mathematics out of the situation. The people that are building the houses are not using mathematics; they're doing it traditionally, as is the farmer. Over a period of centuries there's evolved an optimal way of farming. They farm that way because those have been the farmers that have been successful, and those who have planted at different times haven't been. But we know that there's a reason for this, and that if we can bring out the scientific structure of why it's done, then you can teach science that way. Hence the science and the mathematics comes out of the problem, even though nobody knows it's there. Then you can say, can you do it better?"

Ironically, developing nations have not been prepared for the impact of western science by their western-inspired curriculum. Nebres cites a case involving hybrid agricultural crops in his native country.

"One of the things in the Philippines recently was the effort to produce high-yield varieties of rice. They get much higher yields, but the inputs--fertilizer and pesticides--are also much higher. I remember talking to one farmer who said, 'I know I am getting more rice. I also know I'm getting deeper into debt.' So we had sessions where we taught them bookkeeping, and they began to see that you don't have to wait till the end. You can, with some mathematics, predict what is a break-even point. That's something that, I guess, is very obvious to us, but it wasn't obvious to this man at all."

Many participants in ICME IV stressed the role of mathematics as a culture-derived applications-oriented tool. Several suggested that as other topics became

sufficiently mathematicized, mathematics as a separate
subject might even disappear from the school curriculum.

Problem Solving

The suggestion that the mathematics curriculum
in developing countries be based on the requirements of
the local culture is closely related to the current
call in the United States for greater emphasis on
"problem solving." The problems to be solved derive
from societal needs and the solutions are produced by
application of the pertinent mathematical tools. In
its Agenda for Action [1] the National Council of
Teachers of Mathematics has issued a call that American
mathematics education be founded on problem solving as
its guiding principle:

> The development of problem-solving abil-
> ity should direct the efforts of mathematics
> educators through the next decade. Perform-
> ance in problem solving will measure the
> effectiveness of our personal and national
> possession of mathematical competence.

> Problem solving encompasses a multitude
> of routine and commonplace as well as nonrou-
> tine functions considered to be essential to
> the day-to-day living of every citizen. But
> it must also prepare individuals to deal with
> the special problems they will face in their
> individual careers.

One sees immediately the similarity of these NCTM
recommendations with the remarks of those who urge that
basic primary education be culture-derived and applicable. However, the NCTM agenda noted that the organization does not consider mathematics as solely a tool for
applications:

> This recommendation should not be interpreted
> to mean that the mathematics to be taught is
> solely a function of the particular mathemat-
> ics needed at a given time to solve a given
> problem. ...Each problem cannot be treated

as an isolated example. This recommendation
looks toward the need to solve problems in an
uncertain future as well as here and now [1].

NCTM thus gives notice of its inclination toward
the "heuristic" school of problem solving, which advocates the teaching of generalized techniques for reasoning one's way through all manner of problems.
Heuristic reasoning can claim such distinguished
exponents as Pólya, but its record of actual accomplishment is sparse as well as controversial, largely
ending with Pólya's own clever and fascinating books on
the subject. Heuristics must therefore be carefully
distinguished from applications-oriented mathematics,
although both wear the trappings and use the terminology of "problem solving."

The problem-solving debate infuses the discussion
of universal primary education with generous portions
of misunderstanding as well as disagreement. While
there is a consensus that the failures of the present
western-modeled curriculum have been substantial, the
development of new curricula is mired in contrary
interpretations of basic terminology.

Reference

[1] An Agenda for Action. National Council of Teachers of Mathematics, 1980.

Status of Mathematics Teachers

Donald J. Albers

> I would like to tell you what my candi-
> date is for the problem of mathematics educa-
> tion. The problem of mathematics education
> is: Why is it necessary to have mathematics
> education? Just why is it necessary to teach
> children mathematics? I am not questioning
> the need to know mathematics. Absolutely
> not; that is obvious to me. I am questioning
> why we need to teach mathematics. That is a
> fundamental thing that we don't reflect on
> enough.
>
> -- Seymour Papert

> The system itself will liquidate
> mathematics education as such. Everyone [at
> this Congress] is claiming that the level of
> mathematics education is dropping. Students
> are knowing less and less--everybody claims
> this. What is developing is a feeling, a
> correct feeling I think, that mathematics
> education is a useless appendix in the educa-
> tional system. It's really something that
> does not fit. What's happening is that it is
> being incorporated with other disciplines and
> taught by others.
>
> -- Ubiratan D'Ambrosio

The "new math" storm has passed. The long decline
in SAT scores in mathematics has slowed. Things seem
to be looking up for the beleaguered mathematics
teacher. Nevertheless, the teacher was the focus of
much discussion at ICME IV, for it appears that
vigorous new storms may once again engulf the mathematics teacher.

Declining Status

Papert and D'Ambrosio are referring, in part,
to the impact of technology, particularly calculators
and computers, upon mathematics education. Given that
few teachers have any formal training in the use of
calculators and computers, it is probably unrealistic
to expect them to suddenly incorporate them into their
teaching programs, even if they wanted to. If Papert
and D'Ambrosio are right, then the status of mathematics teachers is likely to decline.

James T. Fey of the University of Maryland commented sadly on the status of teachers in the United
States. "In France it is said: 'The teacher is king.'
I think that's true elsewhere, but it's not true in the
United States. There is a large number of people in
this country who really feel that if they had to, they
could teach in the high school or junior high school
and be very effective at it.

"I have the impression as an outsider that
throughout most of Europe the professional classroom
teacher is respected more in the sense that their professional judgment is taken as word. In Europe there
is much less belief by everyone in society that they
could teach just as well as the professional teacher."

Robert Davis and Thomas A. Romberg recently
visited schools in the U.S.S.R. and in a report on
their visit they commented on the extraordinary position of the teacher (by U.S. standards) in the Soviet
Union:

Hedrick Smith [author of The Russians]
reports on "parents night" in the Soviet
Union. In the United States, "parents night"

is typically an occasion for teachers to
report to parents, not infrequently to be
accosted by parental complaints, requests,
and demands. One might give the capsule
description: parents sit in judgment on
teachers.

"Parents night" in the U.S.S.R. is quite the
opposite. The voice of authority is the
teacher's. If Ivan has not been doing his
homework, this deficiency will be publicly
proclaimed, and Ivan's parents will be admon-
ished and told that they must correct the
problem. There are even reports of authori-
ties restricting the vacations awarded to
parents of children who are doing poorly in
school [1].

Davis and Romberg go on to remark on the warm, caring,
even loving atmosphere to be found in Soviet classrooms.

Gerald Rising of the State University of New York
at Buffalo carries this dismal portrait of American
mathematics education to its logical conclusion. "In
the United States today we face the problem of extinc-
tion of the mathematics teaching tradition. Seven
years ago, I estimated our distribution of teaching
quality as ten percent excellent, seventy percent
pedestrian, and twenty percent unsatisfactory. I
believe that our situation is markedly worse today."

Geoffrey Howson of the University of Southampton
provided little consolation for teachers when he read
to participants at ICME IV excerpts from an article
that he had recently studied:

Perhaps the most decisive criterion we can
take of the real mind of society on the subject of education is the estimation in which
the educators are held. If their work is
really honoured, they will themselves be
honoured for its sake. ...Now, judging by
this test, we are compelled to conclude that
the general estimate of the value of

> education is still very low among us. The
> profession of educator is not honoured.
> ...It is not generally felt, that those who
> are devoted to the office are the greatest
> benefactors of society. Their labours are
> ill remunerated, and often grudgingly. Many
> of them could earn higher wages by a handi-
> craft trade; and most of them would have
> thriven better in the world than they do, if
> they had applied the same amount of mental
> power and activity to the details of busi-
> ness. ...They are not socially recognized as
> equals, by those to whom they may be, intel-
> lectually and morally, far superior. The
> very parents of the children upon whom they
> are conferring, we may say, an intellectual
> existence, consider themselves, generally
> speaking, as the benefactors and patrons of
> the teachers, rather than as benefited by
> them in the persons of their children.
> ...Symptoms such as these are strong evidence
> that the worth of education is not thoroughly
> and heartily appreciated as might, to a
> superficial view, appear to be the case; and
> the want of a deeper sense of dignity and
> value is the greatest obstacle conceivable to
> the real progress of education itself.

This apt yet saddening quotation was written by the
Reverend Edward Higginson in 1839. Thus, a teacher
might conclude that things do not seem to be worse
today, one hundred and forty-one years later. On the
other hand, things may not seem to be much better
either.

Reasons for the Decline

Participants at ICME IV offered numerous
suggestions for the decline in status of teaching.
Peter Damerow of the Max Planck Institut für Bildungs-
forschung thinks that an undue emphasis on objective
testing may increase the distance between pupil and
teacher. "The real reason [that teachers have high
status in Germany] is that the relationship between
teacher and pupil is seen as a very important one in

our whole German system. About 50% of a student's
evaluation is based on the teacher's impression of his
pupil."

Rightly or wrongly, test scores remain very important to Americans, as witnessed by the great concern
over the long decline in SAT scores. Certainly it is
easy for many to reason that SAT scores are declining
because teachers are not doing as well as they did in
the past, i.e., that the teachers are declining. It is
equally easy for the public to attack the practitioners
of "new mathematics," which is now generally regarded
as having been an educational disaster.

Another reason for teachers' decline in status may
be associated with the growth of teacher unions and the
public's belief that professionals should not unionize.
Rising feels that unions have had an additional negative impact on teachers in that they have drained the
teacher's energy and interest from the classroom.

Rising, like Howson, suggests that the image of
mathematics teachers has never been especially good,
and that those who teach prospective teachers have fled
from this responsiblity. "In a modern United States
college or university, as in ICME itself, a mathematics
educator must first overcome the negative image of his
academic home. In the United States we have, for
essentially snobbish reasons, distanced ourselves in
teacher education from the normal school whose sole
goal was to train teachers. This distancing has been
more than just a renaming of our institutions from normal schools to colleges. It has been in resources as
well. Today the undergraduate colleges and universities that provide teacher preparation programs are most
often schools of liberal arts with activities centered
around substantive departments. Education remained or
slid in only at the bottom of this university hierarchy
well below its predecessors of medicine, engineering,
law, and agriculture.

"But worst of all, the faculties of our schools of
higher education have fled from their responsibility
for the preparation of teachers to what is very loosely
designated as educational research, an activity whose

relation to the climb up the professional ladder of the
so-called researcher is inversely proportional to its
relation to classroom realities."

Burned Out Teachers

Another possible reason for declining status
may be linked to the phenomenon called "teacher burn
out." Reports from several investigators point to
signs of growing numbers of discouraged, disappointed,
and frustrated teachers. Louis Smith, in recent case
studies of mathematics teachers, describes the condi-
tion of "burn out" as "a flatness, lack of vitality, a
seeming lack of interest in the curriculum by both the
teacher and the children, a lack of creativity and curricular risk taking, a negativism toward the children-
-they're spoiled, they don't care, they don't try--and
sometimes a negativism toward colleagues, administrators, and college and university training programs
(often decades ago) [2]."

Gerald L. Alexanderson of the University of Santa
Clara recently chaired the joint M.A.A.-N.C.T.M. Committee on the Reported Decline in the Preparation of
Students for Collegiate Mathematics. In their work,
the joint committee interviewed mathematics teachers
from around the country. After concluding a series of
interviews, he is not optimistic.

"In this country teachers are very discouraged,"
said Alexanderson. "They feel that they don't get support from parents, that they're not appreciated, and
that they are not highly thought of in society. And
now the situation is getting much worse. We have fewer
new teachers entering the field, and the teachers who
have been there for a while are getting older and
older. They're getting more and more discouraged.
They feel that it's not as exciting to teach as it was
back in the 1960's. It isn't because society is not
valuing mathematics and science as much as it did then.
In our area the best teachers are being pulled out into
local industry. It seems that the situation in the
classroom is going to get worse! It is getting worse."

It is a fact that teachers are leaving the classroom to pursue other interests, and, remarkably, there
now exists a shortage of mathematics teachers during a
time of declining enrollments. It's quite unlikely
that those who remain in the ranks will have their
spirits raised by the leave-taking of their colleagues.

Howson suggests that part of the problem is due to
the increasing demands on teachers. "I suspect that
some of the excellent teachers who taught me would be
far less successful in a present day comprehensive
school where the demands made on them would be much
greater, even perhaps, excessive. Teaching is now, in
England at any rate, a vastly more complex and demand-
ing task than formerly."

Some of the demands Howson refers to may be found
in the impact of calculators and computers on mathematics teaching. Many teachers have eagerly learned about
these new technological devices and found exciting
applications for them in their classrooms. Many others
have yet to learn about them, much less find applications for them in their teaching. Given the rapid
growth of technological innovation, continuing education for all of these teachers may soon become critical.

Bandwagons

Another related demand of large proportions is
now just beginning to be felt by teachers: a general
call by professional organizations to emphasize applications of mathematics, certainly an about-face from
the "new math" call of the 1950's and 1960's. The
creaky "new math" bandwagon is being replaced by a
shiny new "problem solving" bandwagon.

For example, recommendations 1 and 3 of the recently released N.C.T.M. report An Agenda for Action:
Recommendations for School Mathematics of the 1980's
urge that "problem solving be the focus of school mathematics in the 1980's;" and that "mathematics programs
take full advantage of the power of calculators and
computers at all grade levels." In answer to the obvious question of how to implement these recommendations,

N.C.T.M. also recommends that teacher education programs at all levels should address these matters.

Other professional organizations of college teachers of mathematics have also emphasized the importance
of applications, modeling, and problem solving. As
more and more people get on this new bandwagon, it may
be necessary to oil it, to train additional drivers,
and to find more energy to pull it.

Paul Rosenbloom of Teachers College, Columbia
University, warns us of hidden difficulties in teaching
the applications of mathematics: "A major difficulty is
that in order to discuss properly applications of
mathematics to physics, biology, or economics, one must
know something about these fields. A college teacher,
who is supposed to be a professional scholar, often
feels uncomfortable in discussing subjects he knows
only superficially. How much more difficult is it for
most school teachers, who are not accustomed to studying new topics independently."

Rosenbloom told of experience with sample textbooks produced by SMSG, the School Mathematics Study
Group. "They were tried out with moderate success, at
least when the teachers got inservice education and
consultant help. I do not think any approach to the
teaching of applications in the schools will be successful without a program of inservice education."

There is reason to believe that teachers themselves are very interested in learning about applications. Data from a recent national survey [3] of two-
year college faculty revealed strong interest in applications. The survey group also showed very active
interest in continuing their education. However, they
pointed out, ruefully, that often the application areas
that they were eager to learn more about would be
classed as lower division courses and thus would not
provide them with advancement increments.

Needless to say, most teachers find it difficult
to finance their own continuing education. Many of
them recalled with fondness the National Science Foundation programs of the 1950's and 1960's. Most who had

attended such programs said that these programs had
contributed significantly to improving their teaching.

The institute programs were part of the U.S.
government's response to what was perceived as a technological gap between the U.S. and the Soviet Union, as
manifested by Sputnik. Recent reports on mathematics
and science education in the U.S.S.R. by Izaak Wirzsup
of the University of Chicago and Robert Davis of the
University of Illinois suggest that the Soviet Union
may have launched a very ambitious mathematics program
for virtually all children in the U.S.S.R. The technological implications of such a program might trigger in
the U.S. a new wave of government sponsored continuing
education for mathematics teachers [2, 5].

Federal Intervention in Mathematics Education

The importance of federal involvement in mathematics education was underscored by Marilyn Suydam and
Alan Osborne in their recently completed study [4] of
the causes and effects of mathematics education policy
formation:

> The evidence shows that progress and change
> have resulted from federal intervention.
> Some claim that the federal investment in
> mathematics education has been the vital mar-
> gin in determining whether a change would be
> realized or not. We see little evidence that
> the future will be otherwise. Thus, thought-
> ful policy formulation at the federal level
> is critical since it guides the investment of
> dollars for mathematics education.
>
> It is not sufficient simply to recommend
> increasing the magnitude of the investment in
> mathematics education to make desired
> changes. More money must be invested wisely
> in order to accomplish change expeditiously
> and efficiently in the areas of greatest need
> in mathematics education. Recognition of
> deficiencies in policy formation processes is
> an important first step toward improving the
> payoff of the investment toward improving the

learning and the teaching of mathematics in
the schools.

Suydam and Osborne identify three sources of
failure in the formation of mathematics education policy:
  (1) Educational policy is frequently determined
      without collecting enough information to allow the
      process to be rational.
  (2) Educational policy is frequently constructed
      without using information that is readily avail-
      able.
  (3) Differences in the values held by various groups
      concerned with the schools are frequently not
      recognized in determining priorities.

Elementary Teachers are the Key

       Leon Henkin of the University of California,
one of the organizers of ICME IV, hopes that we will
not repeat the mistakes that he thinks were made with
the introduction of the new math. "Probably the largest mistake made by the mathematicians who started the
'new math' movement in the mid-fifties was that they
thought that by writing books with new topics in them,
they would bring the new ideas immediately and properly
into focus among all the school children. Unfortunately, it didn't work out that way. One of the
major reasons was that the curriculum planners did not
realize the tremendous importance of the teacher in the
learning process. The fact is that teachers' needs are
complex, that teachers' emotions and self-confidence
play a great role in how they are going to present
material."

       Others are concerned that the role of the elemen-
tary teacher is not being addressed with sufficient
intensity with regard to problem solving. Many who
witnessed the failure of "new math" in the schools are
quick to point out that elementary teachers were generally overlooked in the great numbers of federally
funded inservice programs devoted to modern mathematics. "There is no doubt in my mind," said Henkin,

"that the elementary teachers are the key to changing
the teaching of mathematics. The first six years of
mathematical contact for any pupil is with an elementary teacher. Yet for many reasons, the elementary
teacher is often uneasy in presenting mathematics, as
compared with reading or writing or social studies."

In these days of "back to basics" and "problem
solving," it is curious to note that mathematics education requirements for prospective elementary teachers
do not seem to be growing. Some states, in fact, have
reduced their mathematics requirements for elementary
teachers. We now demand that elementary students know
more mathematics. Should not their teachers also know
more?

Several countries, recognizing the importance of
elementary school mathematics, make use of mathematics
specialists in early grades. In the absence of specialists, we must institute inservice programs at all
levels. But as Paul Rosenbloom points out, this will
cost money. "It will be necessary for the mathematical
community to persuade the appropriate agencies to support such a program."

References

[1] Davis, Robert B., et al. An Analysis of Mathemat-
     ics Education in the Union of Soviet Socialist
     Republics. Columbus, Ohio: ERIC, 1979.
[2] Smith, Louis. "Alte." Chapter 3 in Robert E.
     Stake and Jack Easley (Eds.) Case Studies in Sci-
     ence Education. University of Illinois, 1978.
[3] McKelvey, Robert. "The two-year colleges and the
     graduate schools: the teachers' perspective." The
     Two-Year College Mathematics Journal 10 (1979)
     136-142.
[4] Suydam, Marilyn N. and Osborne, Alan. The Status
     of Pre-College Science, Mathematics, and Social
     Science Education: 1955-1975. Volume II:
     Mathematics Education, Executive Summary. The
     Ohio State University, 1977.
[5] Walsh, Epthalia and John. "Crisis in the science
     classroom." Science 80 1 (Sept./Oct. 1980) 17-22.